Keepers of the Trees

Keepers of the Trees

A GUIDE TO RE-GREENING NORTH AMERICA

ANN LINNEA

Skyhorse Publishing

Skyhorse Publishing books may be purchased in bulk at special discounts for sales promotion, corporate gifts, fund-raising, or educational purposes. Special editions can also be created to specifications. For details, contact the Special Sales Department, Skyhorse Publishing, 555 Eighth Avenue, Suite 903, New York, NY 10018 or info@skyhorsepublishing.com.

www.skyhorsepublishing.com

10 9 8 7 6 5 4 3 2 1

Library of Congress Cataloging-in-Publication Data

Linnea, Ann, 1949-
  Keepers of the trees : a guide to re-greening North America / Ann Linnea.
      p. cm.
  Includes bibliographical references.
  ISBN 978-1-61608-007-5
  1. Forests and forestry--Citizen participation. I. Title.
  SD387.C47L56 2010
  333.75'160922--dc22

                                    2009052030

Printed in China

Contents

Foreword

WHEN I WAS A YOUNG graduate student with activist tendencies and a "Stumps Suck" bumper sticker on my ancient Subaru, I kept an old travel trailer on the Olympic Peninsula. Most of my master's thesis on radical environmental philosophy was written there, though I'd take days-long breaks, hiking deep into old growth forests of Douglas fir, Sitka spruce, and western red cedar. This was a solitary period for me, and my time among the trees was a sustaining grace—these trees were animate presences in my life, and I came to know many of them as individuals.

One day I returned to the trailer after a long backpack up the Hoh Valley, still excited about the black bear I'd seen up so very close. There was a large man in suspenders who'd pulled his tiny ramshackle trailer up next to mine. "I see you're tryin' to put me out of business," he said, pointing at my bumper sticker, and laughing. I raised my brows questioningly. "I'm a logger," he told me, naming the large company he worked for. "Oh," was all I could muster off the top of my head. We talked awhile, just chit-chat. But finally he paused. "You know," he looked at me straight and waved his arm toward Hurricane Ridge—the high stretch of Olympic peaks that rose behind the harbor where we camped—"I love these trees as much as you do."

In another hour he knocked at my trailer door, as promised, and handed me a yellow plastic plate brimming with the oysters he'd foraged that afternoon, deep fried into golden, mouthwatering perfection. "Wow, thanks," I managed to say. "I'll return your plate later."

"Oh, don't knock," he said, "I'm going to bed—have to get up at 4 a.m. for work."

Those were the most delicious oysters I've ever had—before or since. They were also the most generously offered, and they turned my idealistic young mind upside down. I didn't stop joining protests against logging and the woeful state of national forest management, but I finally understood, maybe for the first time, what this work was really about. It was not about right and wrong. It was about love. Deep, wild, life-sustaining love, more complicated than I'd ever imagined.

It is easy to love trees. To feel graced by their presence, enriched by their beauty, playful in their limbs. We need trees—a self-evident lesson if you pause to ponder it for even a moment—and Ann Linnea's wonderful Keepers of the Trees will underscore that lesson vividly. But what this book really brings home is the more nuanced lesson—that trees need us. To see them, to plant them, to know them, to "save" them. This saving comes from unexpected places, many of which I hadn't even considered before reading this book. Not just the eco-activists, bless them, but also the pruners, the wood-turners, the storytellers, and yes, the loggers.

The people in Keepers of the Trees are themselves like trees in a forest—coherent, complex, connected, yet individually beautiful. In telling the stories of these varied, amazing people, Ann Linnea gently prods us to consider our own role as tree keepers, to find, in the uniqueness of our individual lives, ways to turn the generous gifts that trees daily bestow upon us back upon these quiet givers.

—**Lyanda Lynn Haupt**
Author of *Crow Planet: Essential Wisdom from the Urban Wilderness*
Pilgrim on the Great Bird Continent
Rare Encounters with Ordinary Birds

Introduction

UP HERE, AT ELEVEN THOUSAND feet, the white, dusty, rocky dolomite soil lies loose on the mountain slopes, swirls around my hiking boots, and kicks up easily in the nearly constant wind. In the mountains, temperature is determined by altitude. It's over 100°F in the basin lands far below but barely 70°F on the trail this July day. Snow and frost can occur in all twelve months of the year in this harsh, exposed place, and the growing season can be as short as forty-five days. Ballooning cumulus clouds indicate that afternoon thunderstorms are already beginning to build around me, so I stay focused on my exploration, knowing I need to be back to my vehicle before lightening begins flashing. Up here, I am a lone moving target for the fire of heaven.

I am dressed in long sleeves and pants, sunglasses and hat. My camera and binoculars bounce on my chest, and my day pack carries the Ten Essentials and an extra bottle of water. I am on a pilgrimage to visit the oldest single living organisms on the planet. They have survived only because they hide out here in this place where few other species can live. They are so rare that only a few groupings of them have ever been discovered—all in remote high mountain

« *Pinus longaeva*

slopes like the White Mountains between California and Nevada and in Great Basin National Park on the border between Nevada and Utah. A few specimens that have adapted to lower elevation are hiding in southern Utah in the national parks there. The species I seek is a tree.

Mountaintop Meeting

Pinus longaeva, the oldest of the three known species of bristlecone pine, are unique, charismatic trees. They live right at or just below the tree line in these few remote mountain sites. Some of the elders stand twenty to thirty feet high, and others grow horizontally along the ground like shrubs trying to stay out of the wind. Bristlecone pines look more like exotic sculptures than trees.

Gnarled, twisted, and able to survive with only a few inches of living bark tissue carrying nutrients between roots and limbs, bristlecones have rock-hard wood that makes them nearly impervious to insect infestation or decay. A two- or three-foot-tall bristlecone sapling that brushes my pant leg may be nearly a century old. We determine the age of trees by counting the annual growth rings they lay down in their trunks. In a dry, harsh year, a growth ring is very thin—sometimes less than a millimeter wide. In a lush, warm year, growth rings in some species of trees can be several centimeters wide. There is no such thing as a lush, warm year for bristlecone pines though. I stop to catch my breath. Around me are living trees that are over four thousand years old and gray sentinels of dead trees up to ten thousand years old. I am walking through a natural time machine.

This forest was here at the beginning of the Bronze Age. The large, twisted tree before me was a sapling when people invented the alphabet, when Babylon was the cultural center of the Middle East, when Stonehenge was erected, when the Chinese were learning to spin silk, and when Asian nomads, who had ventured across the Bering Sea, were gathering nuts and berries in the California valleys below. These trees are supreme survivors—and that's part of why I'm here.

I have always been fascinated by the living ecosystem that holds my life. Though Western science divided the physical world into what it defined as the plant, animal, and mineral kingdoms, twenty-first-century science is discovering the profound interconnection of all things. And with every decade of my adult life, our interdependence with trees has become more completely understood. I am scrambling along this mountain ridge today as an avocational scientist seeking bits of knowledge from a generation of dendrochronologists who have studied the bristlecone pine, as a wilderness educator in awe of the capacity of these trees to withstand their conditions, and as a writer peering at the twists and turns of growth as though I could discern the life stories of these ancient beings.

The first time I hiked in the White Mountains and saw a bristlecone pine, I wanted everyone to see what I saw and to understand what I understood about the significance of these trees. I took dozens of photographs, spent the next winter reading and Googling the Internet for more information, joined several of the tree advocacy groups you will meet in these pages, and became an increasingly active tree advocate. Through this interest, many threads of my professional and personal life began coming together.

As a botanist, I understand that trees are the lungs of the planet. Plants, especially trees because of their size, produce the oxygen on which the rest of life depends. And plants, especially trees, consume carbon dioxide in the process of photosynthesis. This exchange is the foundational symbiosis that supports life on earth. Trees are nature's biggest offering to help stabilize the carbon–oxygen balance. As a wilderness guide and educator, I am concerned that most people take trees for granted and see them as scenery, landscaping, or a harvestable crop that serves as the source of plywood and paper and other necessities. And as a writer, speaker, and listener of story, I know that most people come to understand the natural world through the stories of how that world impacts and changes the lives of other people.

For thousands of years, the bristlecone pines hid out along the mountaintops, and what happened in the human community miles away didn't much affect them. That is no longer true: These trees, and trees everywhere, need people to understand the absolute necessity of protecting them as part of our own survival and the survival of all life. The earth needs her lungs.

So I set off to find people who in one way or another have devoted their lives to trees and whose stories would help us understand the breadth of our relationship with them. I invited people to tell the story of how their lives and their connection to trees grew up together—growth ring by growth ring. And in the middle of their stories, I've spliced interesting bits of science and education to help all of us grow in advocacy, appreciation, and respect for our green neighbors. I call this group of people "the keepers of the trees." There are fourteen stories here, and probably within a mile radius of wherever you are reading this book are a dozen more people whose stories could belong here. And that is the great hope: that thousands, tens of thousands, and hundreds of thousands of people are getting the message about trees and doing something to help.

Finding the Keepers

Writing *Keepers of the Trees*, I interviewed a diverse group of people who represent national, racial, gender, age, and vocational diversity. Over the course of six years, I traveled thousands of miles from Vancouver Island, British Columbia, to the Great Smoky Mountains National Park, North Carolina, and visited urban forests in Los Angeles and Chicago. I was in search of people (most of them would describe themselves as ordinary folk) who had chosen some aspect of relationship to trees as an organizing principle in their lives. It is an honor to hold up their stories of hard work, hopefulness, and creativity. I made their acquaintance through word of mouth—sometimes a chance meeting; sometimes one keeper referred me to the next as I was explaining the idea for this book. Beyond seeking diversity, I followed three criteria:

- someone passionate about their work with trees,
- someone who represents everyone, and
- someone I could visit during my teaching travels.

Most of these keepers are not widely known outside their own communities, and some are almost completely anonymous within their communities. All of them contribute to our scientific, educational, and spiritual understanding of the essential nature of our tree–human interdependence.

No matter who you are or where you live, one of these keepers will speak to your life experience. You will meet the following people in this book:

Andy Lipkis, founder of TreePeople, has been instrumental in ripping up concrete and planting over two million trees in his beloved Los Angeles because he knows trees catch and store precious water.

Ninety-eight-year-old **Merve Wilkinson** supported himself and his family by carefully managing the timber resources on his 136 acres of Canadian forest; yet, there is more standing timber now than when he began harvesting his land sixty-seven years ago.

Learning to be a tree pruner literally saved **Laura Robin**'s life and brought her back to the foothills of her Montana home.

Standing in sawdust and sunlight, **Bud Pearson** has let the art of wood turning soften and transform his Vietnam-ravaged soul.

Great Smoky Mountains National Park forest supervisor **Kris Johnson** is a world expert on the woolly adelgid, an insect ravaging the hemlock forests of the eastern United States.

Will Blozan manages a team of arborists in North Carolina, searches for record-breaking old trees in his spare time, and is passionate about eastern hemlocks.

Shannon Ramsay, founder of Trees Forever, is a genius networker who has everyone from the Iowa Farm Bureau to electric utilities to Syngenta funding programs so thousands of people from Illinois and Iowa can plant trees in cities, towns, and along roadsides and stream edges.

Corella Payne balances a stressful, inner-city Chicago social services job by volunteering her weekends to tend trees in city parks.

As Waverly Light and Power's general manager, **Glenn Cannon** became a spokesperson for the triple whammy investment of trees.

As a wilderness researcher for the U.S. Forest Service and a family owner of a Michigan tree farm, Dr. **John Hendee** shares both a scientific and a spiritual perspective on the importance of trees.

Dr. **Robert Van Pelt**, Renaissance man of the trees, can always be found climbing, drawing, photographing, or studying trees in almost any part of the world.

Cass Turnbull, founder of PlantAmnesty and a charismatic, flamboyant pessimist, is determined to end the senseless torture and mutilation of trees.

Native carver **Russell Beebe** brings a lifetime of experience in understanding how to let the wood teach him to bring the story forward.

After much coaxing from friends and colleagues, I have also included my own story so you might better understand how trees have been my interspecies teachers.

Why Know, Why Now?

In the middle of the technological computer age, the presence and importance of trees in our lives may not be as immediately obvious as it once was. We look around and see metal cars, plastic containers, steel skyscrapers, and manufactured electronics. Certainly there are still things made of wood—paper, furniture, and building materials—but we're starting to use alternative sources for many of these things. We can buy hemp paper, wicker chairs, bamboo flooring, resin decking, and vinyl siding. Do we really need trees?

Yes.

In the West, people still use more wood than any other engineering material. And in developing countries, wood remains essential to daily life at the basic level of fuel for cooking and heating. All life on the planet relies on the precious oxygen produced by trees. Trees have a huge role in mitigating global warming and preserving a sustainable global environment. Numerous scientific studies are currently focusing on calculating exact carbon sequestration values for different types of trees and forests. These studies will radically alter the way that the value of trees is perceived. Our conversation will switch from "board feet of lumber" to "carbon sequestration and oxygen output." In a challenging world, trees are about to be recognized—as Andy Lipkis says—as "our multitasking superheroes."

Saving trees is one of the easiest, most accessible, and often most acceptable forms of activism. This book will teach you how to plant trees for energy conservation as well as for beauty, how to tend the trees already around you, how to advocate in your community for more trees and healthier trees, and how to support the careful use of trees for industry and art. Sometimes advocacy is simple: watching where we put our feet, teaching our children and grandchildren to play gently among the young and old trees that surround them, and emptying our water bottles at the base of a sapling at the end of a hike. Sometimes it means planting trees for Arbor Day, Mother's Day, or Father's Day; getting educated about maintenance; talking with neighbors about the value of trees on the block or along the roadside; and telling stories about why we love them.

Love: we save what we love. That's what I feel on the White Mountains among these elders. Love for these trees and for the tenacity of life they represent. I have come upon them in the middle of their story, carrying my brief life span in my pocket like a small, smooth stone. I am humbled before their longevity and determined to contribute all I can to the dream that a thousand years from now another hiker may stand next to this sapling, now grown and twisted gorgeously in the wind, and ponder this moment when people made the choice to re-green North America and the world.

Join me. Let's start by meeting these keepers of the trees.

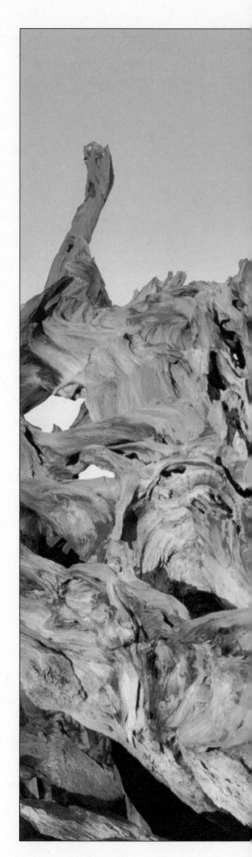

Photo Courtesy: Melinda Kelley

Name: Andy Lipkis

Occupation: Community organizer

Point of Wisdom: Los Angeles can transform itself from the water-squandering capital of the Southwest to the water-saving capital of the Southwest by planting trees. Progress means more trees, not more pavement.

CHAPTER 1

The Tree Man of Los Angeles

Andy Lipkis

So I thought, hey, this is Hollywood.

We can make a storm.

—ANDY LIPKIS

THE LOS ANGELES BASIN is a fragile desert bowl wedged between the Pacific Ocean and the San Bernardino Mountains. There was smog in this valley even before it was heavily populated, a resinous haze of dust and juniper waiting to be cleared by an offshore breeze. Now, the effects of sustaining a metropolitan area of 12.9 million people has created a world-class city that is equally legendary for its entertainment industry, flamboyant lifestyles, clogged freeways, flash floods, fires, and smog. The area may seem destined for decreasing quality of life, but to Andy Lipkis and his organization, TreePeople, Los Angeles is a laboratory for learning how trees transform a densely populated desert environment into a healthy, sustainable city. This is an experiment Andy has been leading all his life.

As a self-confident six-year-old kid, Andy called the *Los Angeles Times* to announce the news of a year-round blooming apple tree in his yard. As a teenager, he organized teams of youths to plant smog-resistant saplings in the foothills around his beloved city. Thirty-five years later, through the same nonprofit he launched to accomplish this first act of urban forestry, Andy and TreePeople have helped plant over two million trees in the greater Los Angeles basin. Still president of that organization, Andy has a strategy for using trees as smog reducers that has expanded to include trees as transformers of L.A. from the water-squandering, concrete capital (nearly half of L.A. is paved) of the American Southwest to a city that retains water for its use.

"A fully mature native oak with a canopy about one hundred feet in diameter has root systems equivalent to an underground water tank and can capture fifty-seven

> **If the trees of Los Angeles gave awards, Andy Lipkis and TreePeople would have a brass plaque that reads as follows:**
> - Andy started the nonprofit organization TreePeople at the age of fifteen to help organize young people to plant smog-resistant trees. Thirty-five years later, he is still president.
> - Andy designed a campaign to plant one million trees in L.A. preceding the 1984 Summer Olympics.
> - Andy colaunched the Urban Greening Initiative Campus Forestry Program after the 1992 Watts riots to educate and employ urban teens.
> - Andy founded T.R.E.E.S. (Transagency Resources for Environmental and Economic Sustainability) to work with government agencies to implement urban and watershed management, including retrofitting an entire 2,700-acre urban watershed.
> - Andy demonstrated that proper landscaping and underground storage could prevent flooding even in record rainfalls and provide a viable water supply even in years of drought.
> - In the extreme droughts and fires of 2007, Andy joined forces with city parks officials to develop a restoration plan for the severely burned regional treasure, Griffith Park.
> - Andy opened the city's Center for Community Forestry at TreePeople's forty-five-acre park headquarters with a LEED (Leadership in Energy and Environmental Design, a nationally accepted benchmark standard for green buildings) Platinum conference center and educational garden.

Primary Components of Smog

1. *Nitrogen oxide:* The source of smog and acid rain, produced from burning fuels such as gasoline and coal.
2. *Particulate matter:* This includes any solid in the air, like smoke or dust, that remains suspended, reduces visibility, and causes breathing problems. It is caused by construction, fireplaces, and industrial processes.
3. *Sulfur dioxide:* Sulfur dioxide is produced by burning coal and paper or smelting metal, and it is a source of smog and acid rain.
4. *Ozone:* Ground-level ozone comes from the breakdown of volatile organic compounds like solvents or as a product of reaction between chemicals produced by burning coal, gasoline, and other fuels. It occurs readily in hot weather and is a pollutant with highly toxic effects on humans and plants.
5. *Volatile organic compounds:* These include carbon-containing compounds ranging from natural substances, such as coal or wood, to industrial chemicals, such as benzene. They are released by evaporation or burning and create hazardous air pollutants.

thousand gallons of water in a twelve-inch rainfall or flash-flood event," explains Andy. "Remove that one tree, and all that water is lost downstream, potentially flooding and damaging homes, neighborhoods, and streets. Furthermore, the water is also lost to recharging local aquifers for use in drier times."

This bit of research is a mantra recited a thousand times—and always with excitement, for Andy was born with a passion for implementing change. Because Andy was born in Los Angeles, much of his understanding about trees has been focused on urban forestry.

Doing Something about Smog

When six-year-old Andy called the newspaper in the early 1960s to report his ever-blossoming apple tree, Los Angeles was one of the smoggiest cities in the world. The desert city bounded by the sea and mountains was already home to 2.5 million people driving cars on the first freeways in the United States. The combination of sheltered bowl, warm temperatures, lots of people, and lots of automobiles made L.A. infamous. "I can remember the air being so bad I'd have to come home from school and breathe steam to get rid of the ache in my lungs," Andy recalls. "To this day, when the light

⌃ As a teenager, Andy founded TreePeople. By the age of 23 he had secured Coldwater Canyon Park as headquarters of the new organization.

changes in a certain way, my respiration gets shallow. I bet millions of people are conditioned to respond the same way."

The six-year-old grew into a fifteen-year-old who attended Jewish summer camp in the San Bernardino National Forest one hundred miles east of Los Angeles at 6,500 feet elevation. "Those forests were a refuge where city kids could run around and breathe deeply."

But even those forests were beginning to feel the effects of smog creeping up from the fast-growing city below. Smog is formed by sulfur dioxide and by the interaction between nitrogen oxides and volatile organic compounds. Rimmed by mountains, Los Angeles holds its pollutants. Add the warmth and UV radiation of California sunshine, and many of the emitted pollutants react with each other to form smog. Smog irritates the eyes, nose, and throat and can damage the lungs as well as the foliage of plants and trees.

In 1970, the year Andy turned fifteen and went to camp, U.S. Forest Service officials announced that smog was killing the dry mountain forests surrounding L.A. at the rate of 10 percent per year. They predicted the complete loss of oak, mountain mahogany, and pine forests surrounding L.A. by 2000. One recommended Forest Service remedy was to plant significant numbers of smog-resistant trees, which in this region included certain species of pines and cedars that do not readily succumb to dangerous pollutants like nitrogen oxides and sulfur dioxide.

"Of course, I had to do something," Andy smiles. "And the perfect place to tackle the smog problem was in the social justice cauldron of camp." After a few days of campaigning with camp management, Andy mobilized fellow campers and staff to pull up an old parking lot and plant it with Coulter pines. "It was a start," he explains. "It helped that I had worked with my parents on Gene McCarthy's presidential campaign as a twelve- and thirteen-year-old. I understood how to take ideas and put them into action one piece at a time. It was an incredible thing to transform that piece of earth and have such immediate gratification."

Returning to the city to eleventh grade, Andy was ready for action. "All I wanted to do was get kids into the mountains to experience their power and taste what's possible. I wanted them to get away from thinking they had to wait, go to school, and get a job before they could do anything important with their lives."

With the help of a teacher who believed in him, Andy enrolled in an experimental high school that required a community service project. His project was to build a summer camp in the mountains solely for the purpose of getting kids to plant trees. "For three years I tried, failed, quit, learned, and started again. Every time I quit I was able to recharge my batteries. All of the work built a huge compost pile of lessons and possibility. Failure never is final."

As Andy shoveled his way through trial and error, the rich compost finally germinated a feasible plan: use existing camps and encourage each camp to plant one thousand trees to restore the environment. Next problem: finding money to buy twenty thousand smog-tolerant trees being held in a state nursery that was going to plow them under if the trees were not purchased.

"So again I called the *L.A. Times*. I think it is part of my DNA to insist that the news media help public causes," he grins. On April 23, 1973, the *Times* ran a front-page story about the kid who wanted to rescue trees. The headline read, "Andy vs. the Bureaucratic Deadwood." The next day sacks of mail began arriving at Andy's house. School kids, seniors, and families all expressed gratitude that someone was doing something for the dying forests. Within three weeks he had received over $10,000—mostly in coins and dollar bills.

"We put the money in the bank. An older cousin helped me file incorporation papers. American Motors donated Jeeps to move the trees to the camps. And TreePeople was born."

So, without an advanced degree and in an era when the phrase "L.A. urban forester" seemed more like a non sequitur than a profession, Andy had inadvertently become one of the first of his kind. Urban foresters work to create forests that thrive within the city. They must love both trees and people, for their task is to watch the health of individual trees while figuring out ways to improve and sustain the health of the city's ecosystem. Generally urban foresters have degrees in biology or biological processes, but through TreePeople, Andy had the wisdom to recruit people with the education and technical skills to bring urban forestry to the forefront of L.A. city management thinking.

A recent study by American Forests showed that 448 of the United States's largest urban areas lost more than 3.5 billion trees in the past ten years. Andy has been a relentless pioneer in educating Los Angeles's citizens and leaders to the full beauty of urban trees. And although the organization has planted over two million trees from San Bernardino to Santa Barbara since its 1973 beginning, the tree canopy of Los Angeles is still just 18 percent—less than the canopy of New York City.

Another Flamboyant Project

The organization's first real growth spurt occurred in 1981 when twenty-six-year-old Andy responded to a city air-quality management plan that estimated the city could meet requirements of the 1970 Clean Air Act by spending $200 million to plant a million trees in twenty years. By 1970 and the passage of the Clean Air Act, the role of trees as a natural antidote to certain kinds of air pollution was well established.

TreePeople approached the city council and declared it could save the city $200 million because it could organize the public to do the job in three years (not twenty)—in time for the 1984 Summer Olympics—and at no cost to the city. The challenge the young organization took on was to "inform, enroll, guide, and inspire the public on an ongoing basis without a public relations or advertising budget."

With the pro bono help of an advertising agency, a thirty-second TV spot featuring Gregory Peck, a radio commercial, and artwork for ads and billboards, TreePeople launched the "Urban Releaf" campaign. The successful project had moments of publicity genius: getting eight, forty-foot National Guard trucks to deliver one hundred thousand donated seedlings; enrolling a local TV station to develop a "treemometer" to measure the twenty-four-hour hotline registrations of planted trees; and using General Telephone's Urban Forest Ranger program that involved seventy thousand kids.

Three years later, just four days before the lighting of the Olympic torch, the millionth tree—a flowering apricot—was planted in Canoga Park. Though outwardly the campaign was a huge success, in typical analytical style Andy cautions other organizations about getting caught in "bean counting." "TreePeople

Photo Courtesy: TreePeople

« *To reach their million tree goal in time for the 1984 Olympics, TreePeople relied on a number of grassroots marketing tactics including bumper stickers and license plates.*

Even before the million tree campaign, Andy and TreePeople were capturing city attention. In 1980 three thousand TreePeople volunteers assisted local homeowners after excessive rains and mudslides. Andy appeared on The Tonight Show. Host Johnny Carson made a personal contribution to replace tools lost during the relief work. »

is still apprehensive about a repeat of a campaign for Los Angeles that hangs success on a number . . . you'll go crazy getting to that number. You'll beg people to let you count their trees. They'll ask if you want to count their rose bushes and vines, too." The lesson learned in the million-tree campaign was that the real job of tree planters is to plant carefully and engage in long-term care. Survival rate is more important than initial numbers. The Olympics came and went, the United States took home eighty-three gold medals, and the social and environmental challenges of the L.A. metropolitan area remained as overheated as the valley itself.

The Watts Riots and the Next Step for TreePeople

Andy found himself deeply disturbed after the civil unrest in the L.A. neighborhood of Watts in 1992. Twice in Andy's lifetime Watts had gone up in smoke. In 1965 this residential section of South-Central Los Angeles had been the site of six days of race riots that claimed thirty-four lives and caused over $200 million in property damage. When race riots erupted again in 1992 after the acquittal of white police officers who were videotaped beating a black motorist, fifty-eight people died, and $1 billion in property was destroyed.

Andy's city was in trouble again. "TreePeople had been saying that trees build community, but the riots were a clear sign to me that we weren't getting the job done. I knew one of the real core issues was lack of jobs, because in 1990, 40 percent of African Americans living in Watts were living below the poverty line."

Firm in his resolve to respond to social issues through environmental channels, Andy worked to co-create the Urban Greening Initiative of the U.S. Forest Service. This initiative brought $2.5 million in job programs and urban forestry projects to Watts and other areas of South Los Angeles.

Andy is the ultimate modern environmentalist. As soon as progress is made in the solving of one problem, he once again steps back to survey the larger picture. Andy's brilliant mind thinks in terms of whole ecosystems. Once the Urban Greening Initiative was in place, Andy began thinking about the facts, as he knew them:

- People in South L.A. needed jobs.
- Recent news reports had made it clear that huge amounts of federal dollars were available, specifically for increasing the size of L.A.'s storm drains.
- Storm drains meant usable water was being flushed away.
- Statewide, Los Angeles had a reputation of being a water thief—importing 85 percent of its water.

"I remember the night this information began to click around in my mind like pieces of a puzzle," he says. "I was driving home from a TreePeople board meeting. It occurred to me that probably no one was taking responsibility for looking at Los Angeles as a whole ecosystem. If we were importing 85 percent of our water and spending billions to create more and more concrete tunnels to funnel away the fifteen inches of rainfall we do receive a year, something was seriously wrong. It made no sense."

The boy who had persisted in his belief that good ideas should be manifested into action became a man who thought in nontraditional patterns and knew how to get things done. Andy's initial strategy to transform L.A. from water squanderer to water saver was to get agencies talking to each another. However, as he set an intention to launch communication between L.A.'s largest water agencies, the Army Corps of Engineers and the L.A. County Public Works announced plans to increase the size of water diversion aqueducts. Their plan proposed raising the concrete walls of the Los Angeles River another four feet along the river's last twelve miles to protect against a potential hundred-year flood. (Encasing the Los Angeles River in concrete began in the 1920s in response to a major flood.) TreePeople, which generally works with public agencies, not against them, joined a 1995 lawsuit against this half-billion-dollar "Band-Aid" approach to water management.

The lawsuit failed, but it succeeded in launching the dialogue Andy envisioned. One of the provisions of the settlement was that the County Department of Public Works investigate alternatives to traditional storm-water runoff. Another result was the formation of the T.R.E.E.S. project.

T.R.E.E.S., or Transagency Resources for Environmental and Economic Sustainability, became the vehicle for implementing Andy's vision for winter water retention and summer irrigation. The underlying mission of the project was to demonstrate that local government agencies could work together in ways that saved money, created jobs, and addressed pressing environmental problems. All parties agreed that the place to start was the collection of facts.

TreePeople created a mock storm at a single-family dwelling in South Los Angeles. Note dignitaries at left watching with umbrellas. The property captured nearly all the "rain" by using a cistern, vegetated swale retention grading, and a driveway grate and drywell. »

In 1997, the U.S. Forest Service and T.R.E.E.S. gathered engineers, landscape architects, city planners, urban foresters, and public agency staff for a four-day design process. The challenge for this group was to retrofit all of Los Angeles for sustainability. They were to draw up landscape plans for five typical urban sites: a single-family home, a multi-family complex, a commercial/retail center, an industrial site, and a public high school. The scenario was to reduce water importation by half while not losing any water during a 130-year flood event.

The planning process produced carefully documented designs demonstrating best management practices with a goal of restoring life to the Los Angeles River, turning storm water into an asset, and reducing water importation by half. However, the two previous parties in the lawsuit, the Army Corps of Engineers and L.A. County Public Works, were not impressed by all the theories.

A Storm Made in Hollywood

Undaunted, T.R.E.E.S. (whose funders included the City of L.A. Stormwater Management Division, L.A. Department of Water and Power, U.S. Forest Service, U.S. EPA, City of Santa Monica, and Metropolitan Water District) began turning some of those designs into a real-life demonstration. Around a twelve-hundred-square-foot house in an underserved area of the city, the T.R.E.E.S. team installed cisterns, graded

and bermed the lawn as a retention pond, planted trees, and applied mulch. It was August 1998, months away from any winter storm that might prove or disprove the theory of water catchment.

"So I thought, hey, this is Hollywood," Andy explains excitedly. "We can make a storm." With a rented water truck, T.R.E.E.S. created a flash flood by dumping four thousand gallons of water on the house in ten minutes.

"It was quite a media event," explains Andy, gesturing excitedly. "A helicopter full of TV cameras documented that none of that water ran off into the street. It was a life-changing experience for a number of city and county officials, including Carl Blum, deputy director for L.A. County Public Works."

Blum had spearheaded the project with the Army Corps of Engineers to add four feet of concrete to the lower twelve miles of the Los Angeles River conduit that TreePeople and other environmental groups had taken to court. After standing among the public officials witnessing the house dousing, Blum went back to his office and had his engineers check out the numbers T.R.E.E.S. gave out at its press conference.

"This was such a profound moment for me," says Andy. "Here is the authority, the icon on the other side. He called up the next morning."

Blum said, "I'm calling to say I'm sorry. I did not understand before. We think you've cracked it. We need to take this single-house model and see if it will work on a larger scale as fast as we can."

Blum immediately decided to put on hold a $42 million project to install conventional storm drains in Sun Valley, located northwest of downtown Los Angeles. He ordered his staff to conduct a feasibility assessment of how to incorporate the storm-water best management practices considered in a T.R.E.E.S.'s model to transform the highly flood-prone subwatershed into a model watershed where rainfall is captured, recycled, or stored.

Transforming a Seemingly Hopeless Situation

Ever since development paved the Sun Valley stretch of the San Fernando Valley in the northwest greater L.A. basin in the late 1960s, even ordinary winter rainstorms had forced children and adults to wade knee-deep across roads and major thoroughfares turned into rivers capable of stopping vehicles. Streets had been built without storm drains to whisk water away, and promises to fix the problem had never been fulfilled.

Blum's decision to apply the T.R.E.E.S. solution to the problem launched a whole series of events, including the formation of the Sun Valley Watershed Stakeholders Group. Composed of politicians, environmentalists, government agencies, and community groups, the stakeholders began envisioning and creating options for a community in L.A. previously dubbed "the gravel pit and junkyard district of Los Angeles." Landscaping, installing a network of filtration systems, using gravel pits as water-retention systems, retrofitting homes and businesses, creating parks, and un-paving school yards were all part of the

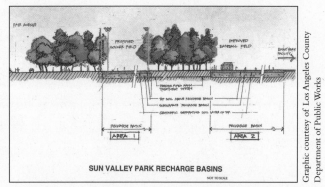

⌃ *Schematic drawing of the Sun Valley Park project*

plan to create an urban watershed that worked. For two and a half years, Blum's engineers worked on design of the system and cost-benefit analyses/computer models.

"The boss [Carl Blum] had had an epiphany," explains Andy. "What we didn't know was whether or not his staff would get it, too, but they have. Essentially what happened is that we are now putting in an urban forest instead of a storm drain in a 2,700-acre area with eight thousand homes."

In his exuberance about the apparent openness of some government officials to begin seeing the potential for changing the "grab and waste" philosophy that has pervaded Los Angeles for decades, Andy does not hesitate to explain that the initial investment cost more than a traditional solution:

> The cost of a storm-water drain for Sun Valley was going to be $42 million. Installing an alternative system of storm-water management may eventually exceed $180 million. But you are comparing apples and oranges. The $42 million gets rid of floodwater by dumping it, and all the pollutants it gathers, into the ocean. The proposed alternative system will produce $100 million worth of water-supply benefits. It will produce sanitation savings of $40 to $80 million because green waste that has been picked up by the city will now be used to mulch and hold water. Its benefits, including energy savings, add up to more than the costs.

In 2007, the storm-water portion of the project was completed (Sun Valley Park), enabling an average of ten million gallons of water per year to soak into the ground to recharge the existing aquifer. As part of that project, three underground units were installed to remove garbage, heavy metals, sediment, oil, and grease before water flows to two large basins that gradually release clean water back to the earth.

While the Sun Valley project was being built, Los Angeles experienced the year of the 130-year flood. In the winter of 2005, over thirty inches of rain fell in a three-month period, and one of TreePeople's projects of working with governmental agencies survived a supreme test. Days before the rains began, an elementary school in Westchester near Los Angeles International Airport finished a retrofit of its playground that included an 110,000-gallon underground cistern, removal of 40 percent of the asphalt, and the planting of grass, gardens, and 150 trees. That winter an estimated eighty-two billion gallons of runoff—including oil, trash, and other pollutants—poured through the city into Santa Monica Bay, but no flooding occurred at the Westchester elementary school. "Although the cistern overflowed into storm drains, the pretreatment system had taken out all the trash, oil, and garbage," said Principal Robert Burke. And the next July, with temperatures in the one hundreds, the elementary school had its own underground reservoir to draw from to keep its new plantings green and healthy.

Conditions Keep Changing

While 2005 was the year of the big floods for Los Angeles, 2006 and 2007 were the years of extreme drought and raging fires. TreePeople, with its track record of thirty-plus years of organizing citizens and caring for the environment of Los Angeles, stepped forward to help. TreePeople, working with L.A. City Recreation and Parks staff, prescribed the best restoration plan for severely burned Griffith Park, the largest municipal park with urban wilderness areas in the United States.

TreePeople has been working with forest fire reclamation for decades. It has a long, strong history of working with the U.S. Forest Service, beginning with bringing volunteers to the Angeles and San Bernardino national forests in the 1970s in response to smog and forest fires. Its first big planting of trees occurred in 1975 after the Big Tujunga fires of 1975 that burned portions of Big and Little Tujunga Canyons, La Crescenta, La Canada, and all the way to Mt. Gleason. TreePeople's planting efforts from those fires continued through the end of the 1970s.

Because of the recent increase of fires in Southern California, TreePeople's fire restoration work will also be increasing. In addition to the Griffith Park work, other work includes the planting of twenty thousand seedlings in areas of the San Bernardino National Forest that burned in the 2003 fires and ongoing work to bring teams of volunteers to the Angeles National Forest affected by the tragic fires of 2009. In June 2008, TreePeople received a $1 million grant from the Boeing Company, followed quickly by a $1.5 million grant from the Walt Disney Company, to launch a comprehensive California Wildfire Restoration Initiative built around public–private partnerships. Over the next five years, the initiative will train and support at least 7,500

⌃ *Andy and thousands of volunteers work to plant trees after fires and floods and as part of water retention for the City of Angels.*

volunteers to restore over ten thousand strategic acres, targeting forest and woodland areas that burned so intensely that recovery on their own is unlikely.

Drought and flooding are both conditions that are forecast to increase with climate change. Bringing people together to restore the forest watershed by strategically planting trees both in the mountains and on city streets is increasingly seen as a central part of the solution if we are to successfully adapt to what the future holds. Andy calls trees "our multitasking superhero partners" in making communities more livable and resilient. The work and vision of TreePeople remains to help nature heal Los Angeles. The group offers sustainable solutions to urban ecosystem problems by focusing on three areas:

1. training and supporting communities to plant and care for trees,
2. educating schoolchildren and adults about the environment, and
3. working with governmental agencies on critical water issues.

Andy is the first to say, "While all of this large, visionary work about watersheds is going on, we keep educating children and adults, and always, always we keep planting and caring for trees. The true work and credit for TreePeople successes comes from our volunteers. We have one of the nation's largest environmental education programs that connects communities and families to the nature that surrounds them." Over three hundred schools in L.A. County participate in TreePeople programs—that's over two hundred thousand students each year, with ten thousand kids coming to headquarters in Coldwater Canyon Park for ecotours.

TreePeople's headquarters in Los Angeles's Coldwater Canyon Park had previously consisted of a former fire station on a forty-five-acre ravine overlooking the San Fernando Valley forested with oaks, black walnut, and native trees. For more than three decades, the organization has cared for the L.A. environment out of this modest facility, with most of its public work happening away from headquarters out and about in L.A. County. However, in January 2003, the group broke ground for the Center for Community Forestry at Coldwater Canyon Park, the new headquarters of TreePeople. In October 2008, the entire campus was completed, with climate-tolerant landscaping and a sustainably designed parking lot (draining, of course, into a 216,000-gallon cistern for irrigating the grounds). New construction includes a LEED (Leadership in Energy and Environment Design, a nationally accepted benchmark standard for green buildings) Platinum-certified conference center seating 150 people, an education center, a watershed teaching garden, and a nursery.

⌄ *Every year more than 10,000 elementary students visit TreePeople's headquarters to learn about their urban forest.*

Cites are currently designed to separate water supply, stormwater pollution, sanitation, energy use, air quality, and more. This disintegrated approach is costly and also wastes valuable natural and human resources. »

LOS ANGELES TODAY

Current approach causes multiple problems

TOO MUCH GREEN WASTE

Sanitation: Trucking to Limited Landfills

TOO MUCH WATER WASTED

Stormwater Pollution: Clean-up & Prevention

TOO MUCH ENERGY DEMAND

Air Quality & CO₂ Mitigation: Conservation Measures

TOO LITTLE WATER

Water Supply: Seeking New Sources

TOO MUCH WATER

Flood Control: More Concrete

TOO FEW JOBS

Economy: Chronic Unemployment

Dis-integrated approach wastes resources, duplicates efforts and imposes unsustainable practices.

Much about the new TreePeople headquarters is educational: The buildings use recycled and non-toxic materials, water-conserving plumbing, and energy-efficient designs that minimize the need for heating and air conditioning. While many of these green ways of building are common in other parts of the country, they are still fairly new to Los Angeles. Countertops in the new center are made from pressed sunflower seeds, and insulation is shredded cotton fiber from discarded blue jeans.

Andy has been fighting for his beliefs for so long, it is almost second nature for him to assume an uphill battle for his ideas. Here is a man who started and stopped, started and stopped, several times before launching his first really big idea—organizing youth to plant smog-tolerant trees. "All I wanted to do was be normal," says Andy, remarking from the vantage point of years and graying hair. "I kept thinking if I just enrolled in college and stuck with something, I could have a normal career."

Normal is defined as "usual, typical, or expected—conforming to a standard." Instead of "conforming to the standard" of environmental conscience in modern L.A., Andy has methodically been rewriting that standard. At five feet, nine inches tall and wearing a TreePeople jacket, Andy does not appear to be an intimidating figure.

LOS ANGELES POTENTIAL

An integrated approach creates multiple solutions

Sanitation Mulching

New Water Source

Reduces Solid Waste

Air Quality & CO₂ Mitigation Tree Planting

Watershed Management

Improves Water Quality

JOBS

Reduces Water Demand

Rainwater Capture & Reuse

Reduces Run-off

Flood Prevention

Integrated approach also creates jobs and liberates funds for emerging green technologies.

But the clarity of his mind, the thoroughness of his preparations, and the impressiveness of his track record make him a bit of a multitasking superhero himself, a force public officials respect. And his teachers have always been trees. "We can and must learn from trees," says Andy. "Progress means more trees, not more pavement."

« *Another approach is to manage our urban environment as a functioning community forest. This creates multiple solutions for critical problems and also releases funds for green collar jobs and technologies such as rainwater harvesting, tree planting, and renewable energy.*

Photo Courtesy: Ann Linnea

Name: Merve Wilkinson

Occupation: Woodlot owner

Point of Wisdom: If a forest is selectively and sustainably logged, it can produce forever. Clear-cutting forests and slash burning destroy them for future generations.

CHAPTER 2

The Steward of Wildwood

Merve Wilkinson

The product coming out of the forest must take second place. The welfare of the forest must come first.

—MERVE WILKINSON

A MAN WHO HAS WORKED HARD into his nineties, Merve Wilkinson resembles the old-growth trees he has spent a lifetime protecting. His hands are gnarled from years of holding chain saws. His craggy posture is stooped. And like an old tree, he continues to nurture all around him. If Canada manages to save a significant portion of its remaining old-growth western forests, it will be in part a credit to Merve's hard work and advocacy.

Merve is a product of the forest. From infancy on he has lived on a tiny section of northern temperate rain forest on Vancouver Island, the largest island off the coast of western North America. The rugged Pacific Ocean side of the island with its deep fjords and mountains receives the greatest precipitation in North America—ten to twenty feet of rain per year. This wild coast is still home to some of the last great stands of old-growth coniferous forests—western red cedars, Sitka spruce, and western hemlock that tower over 180 feet tall and attain diameters so large that a dozen people joining hands can barely encircle them. Except for California's giant sequoias and redwoods, these are the largest trees in the world. Merve's home, which he calls Wildwood, lies nestled on the more settled eastern coast of the island. Here average rainfall amounts are considerably less, so the trees are smaller but still huge by any standard.

In 1915, when Merve was four years old, his parents moved from a coal mining camp in the growing city of Nanaimo on Vancouver Island's eastern shore ten miles down a dirt path to large acreage on Quennell Lake. There they learned to eke out a living with other first-generation immigrants from northern Europe. When the stock market crashed in 1929, Merve was eighteen years old and already a hardworking member of his community. Growing to manhood during the Depression, the handsome, energetic Merve found jobs up and down the island laying phone lines, building fireplaces, and running pulp through the pulp mill in Powell River. He saved enough money to marry a young schoolteacher, Mary Carpenter. On their combined incomes, they were able to purchase acreage on Quennell Lake adjacent to Merve's parents' land for $1,500. It was 1938, Merve was twenty-seven, and the young couple owned 136 acres of prime, forested land, which they named Wildwood. Merve was home. He would never move again.

Merve and the forest quietly, patiently sculpted one another. He studied each tree, watched the woodpeckers and squirrels, observed where the water pooled in the rainy season, and noted the insects on and in the soil. He was determined to sustain himself and Mary on the land and to sustain the land itself. Someone else would have clear-cut the forest and gone into farming; Merve cut only limited numbers of trees every few years and let the forest teach him how to tend it as a renewing resource. Because of this approach, the forest retained its natural composition—a good balance of large and small trees, wildlife, and shrubs. Over many years the composition was also affected by the exclusion of fire, which changed the forest from one dominated by large Douglas fir trees to a more closed canopy.

Following the tradition of modern European forestry, Merve knew he had to harvest carefully so there would always be wood in his forest. This kind of thinking was an anomaly in western North America, where logging was king, high grading and then clear-cutting were the rule, and forests seemed endless. Though he could not know it at the time, Merve and the Wildwood forest were preparing to become a showcase for sustainable logging. When Merve turned seventy-three, his efforts gained national prominence.

In 1986, the energetic septuagenarian, who was still doing all his own felling, yarding, and selling, was watching reporter Cam Cathcart's weekly *Pacific Report* on the Canadian Broadcasting Company. "The mayor of Ft. Nelson was on, and he was wailing about how the timber in his area was disappearing," explains the lively, confident Merve. After the program, Merve phoned Cathcart.

"Mr. Cathcart, I'm very interested in the show you just aired. Does it ever occur to the mayor of Ft. Nelson that forestry doesn't have to be done in a way that devastates and destroys the countryside?"

"What makes you say that?" asked Cathcart.

"Well, I've been caring for a piece of property since 1938 by selectively and sustainably logging, and I've got more timber now than when I started." Within days Cathcart's film crew was roaming Merve's Wildwood property to see if this could possibly be true.

Driving down the one lane dirt road into Merve's picturesque property,
the film crew found themselves transported back in time. ⌄

Photo Courtesy: Ann Linnea

Driving down the one-lane dirt road into Merve's picturesque forest property on Quennell Lake, the camera crew found themselves transported back in time. Merve was waiting to welcome them on the doorstep of the log home he had built and crafted out of the wood from his own land. Nearby was a small barn for animals, a chicken coop, and a sizeable vegetable garden. It was not the homestead, though, that captured the attention of the camera crew. Their eyes were drawn to the trees that were everywhere—large and small all mixed together. Accustomed to seeing even-aged stands of twelve- to twenty-inch-diameter Douglas firs that resulted from replanting clear-cuts, the crew panned their cameras down into moss-covered swales where old western red cedars, with their feathery branches, towered over small clumps of young red alder trees that boasted a healthy understory of native shrubs like salmonberry, huckleberry, and the ubiquitous evergreen shrub salal. Atop the forested slopes above Merve's cabin, a bald eagle was sitting in one of several five- to six-hundred-year-old, never-cut Douglas firs. Below these giants, mixed-age stands of western hemlocks and Douglas firs were striving to become the next old growth. And everywhere there was wildlife—hummingbirds, wrens, warblers, deer, raccoons, and even the occasional bear and cougar. To the untrained eye, it appeared as if no logging had ever occurred here.

Merve gave Cathcart's crew the lecture he had shared with many of his local friends who remained skeptical of his unorthodox logging practices. He likened his woodlot to a tree garden that he harvested every five years. "I have mental maps of every tree on the property. I decide which trees need to stay until they reach optimal cutting size, which are ready to cut, and which need to be cut for disease or thinning. In those years that I do cut, I take out about 340,000 board feet and sell it to local lumber mills. When I purchased this property in 1938, a timber cruiser estimated it contained 1.5 million board feet. In another three years, I will have extracted 1.4 million board feet. In other words, in the first fifty years, I will have effectively cut the whole thing down—yet there it stands."

Cathcart aired the segment about Wildwood a few weeks later. Almost immediately Merve's profession shifted from that of private forester and logger to public educator and activist. Foresters, university students, and interested citizens from around the world began arriving on Merve's doorstep to study the phenomenon of a forest that had been cut and not destroyed.

Merve carefully studied his land right from the beginning, cutting only those trees from areas of good growth rates. »

Younger Years

From the very beginning, Merve was encouraged to succumb to the prevailing wisdom—cut the trees and take the money. Immediately after he purchased the land in 1938, he had been offered the purchase price ($1,500) in exchange for clear-cutting all his timber. The money would have given him and Mary funds to fix up the crude cabin included in their purchase, but Merve didn't hesitate. "No way would I clear-cut it. I grew up here. I loved the forest." He responded out of love and instinct, and in that moment he recognized the need for knowledge, too. He wanted to understand how the land could support his family and still retain its beauty.

So, when their son, Denis, was just a year old, Merve and Mary took a three-week youth training program offered by the provincial government and the University of British Columbia. The Great Depression was still in full swing, with breadlines snaking down the streets of Canada's largest cities. This training program was one example of how the Canadian government was working to help its citizens. Merve and Mary learned about horticulture, pruning, the care of poultry, farm mechanics, livestock raising, and blacksmithing. They were eager to begin to make a living from their land.

Photo Courtesy: Ann Linnea

Since Merve and Mary did well in their youth training program, they were invited to apply for an eight-week advanced course in Vancouver at the University of British Columbia campus. They were accepted and offered a cabin for housing. This education marked the turning point: Merve moved from someone who loved Wildwood to someone who would begin to understand how to be its steward. And Dr. Paul A. Boving was instrumental in that transformation. When he discovered that Merve and Mary owned 136 acres of old-growth timber, he said, "Good Lord, Merve, you should be taking forestry, not agriculture. And I don't recommend any of the classes being taught on the American continent."

Merve certainly didn't have the money to study abroad. However, a few weeks later, Dr. Boving motioned to Merve to stay after class. He explained that he would tutor Merve in the principles of European forestry. Remembering the event as if it were yesterday, Merve has a strong storytelling voice: "It took me about three years to do the reading and paperwork. During that time, I was back on the farm, cleaning up downed timber and selling it for firewood while studying and doing the forestry exercises on my own property. Mary was teaching school. We were beginning to create our life at Wildwood."

Merve has always been deeply integrated into the life of his remote community. In those years his neighborhood, Yellowpoint, had only a few, poor-quality roads, and travel was often by rowboat. Yet like many rural areas, there were numerous gatherings to share food and music. Merve helped establish a co-op and credit union. There was an activity club and local music and theater groups. Merve loved the local life, from campaigning for local politicians to teaching his son and other boys to do chores and drive his old truck. And like most of his neighbors, he did many odd jobs to make ends meet—building and repair work with both wood and masonry, in addition to working with his trees.

Just as a gardener must make elaborate preparation for the first crop—tilling, planting, and mulching—Merve, the young forester, had to make extensive preparation to properly harvest his first crop. He had graphed his land and assessed potential production in each area as poor, intermediate, or good based on growth rates. His patience gathering the information his forestry course recommended was impressive, especially in the face of pressure from folks in the logging industry around him. In a voice as clear and resonant as when he was hearing this admonition, Merve can still mimic his doubters: "Hell, you can't take that kind of time, you'll go broke." Telling the story sixty years later, he laughs. "I just told them, we'll see." When he tallied up the scale slips from that first selective cut in 1945, he discovered he had made money on the operation. "I had success right off the bat. It paid for my own time for felling; paid for the yarding and the hauling; even paid for my time figuring the cut—all of that on prices not yet recovered from the Depression."

Life Lessons

Living in a community where economic stability was an ongoing issue for everyone, Merve and Mary began to feel confident that the land could support their small family. Mary was as recognized for her homemaking skills—cooking, sewing, and gardening—as Merve was for his handyman skills. The end of World War II was an expansive and optimistic time for most everyone in the Yellowpoint community. The postwar optimism of victory in Europe and Asia trickled down to even those on a subsistence income, but the good energy began to be marred by reports of mysterious headaches and even paralysis in the community.

Years later the Canadian government would admit that radioactive rain had reached British Columbia and Alberta from Hiroshima and Nagasaki, but at the time Merve and Mary's community was left to unravel the mystery of strange illnesses appearing in too many neighbors. One of Merve and Mary's neighbors, Jim Galloway, who had suffered from severe headaches, requested that an autopsy be done when he died. The results were conclusive: a massive brain tumor commonly caused by exposure to radiation. In 1949, Mary began to suffer severe headaches herself. "Denis was 13 years old," Merve says, as he remembers a time he would like to forget. "I don't like to talk about that time. Bad things happen, and you must find your way through them."

After a few months of headaches, which she treated with aspirin, a cup of tea, or rest, Mary went to a doctor. At first he prescribed sedatives, but soon they no longer worked. Meanwhile, three of their closest friends began experiencing headaches. Two were diagnosed with brain tumors and survived surgery but were left with little quality of life. Mary's doctor recommended that she go to Vancouver for a brain scan. The week before she was to go, she wrote a careful note saying how much she loved Merve and Denis and explaining that she believed she had a brain tumor and could not tolerate what she saw coming. She walked into the woods one night while Merve was gone to the neighborhood activity club and shot herself in the head with the family shotgun.

Merve was devastated. He didn't know how to cope except to care for his son and tend his forest. Neighbors and friends worked hard tending to both of them. Grace Richards, who had helped Mary around the house with the considerable chores of

High-grading versus Selective, Sustainable Cutting

High-grading means cutting the best and leaving the rest—a common practice in Canada in the past century. It describes selective forest harvesting that removes the most valuable trees, leaving behind trees of poor form, health, value, or potential. High-grading delivers short-term economic gain—for landowners, loggers, and the sawmill—and generally long-term devastation for a given piece of land.

Merve does *selective, sustainable cutting*—cutting some trees in a stand with the intention of maintaining a healthy forest. Selective, sustainable cutting takes some of the best and leaves lots of the best.

Too often the trees left after high-grading are prone to weather and pest damage, and the dominant species is then completely removed. With much care a high-graded forest can return to a respectable timber stand. Usually, though, it is ignored and eventually clear-cut for pulp.

the large garden and farm animals, was especially helpful to the bereaved father and son. Like a tree struck by lightning, Merve struggled to regain his life force. The incident deepened his basic distrust of authority.

Four years later, in 1953, Merve had recovered enough to travel to Europe for the first time. He enjoyed seeing the countryside, meeting people, and visiting the places from which his parents and grandparents had emigrated. He saw the examples of good forestry practice that he had been following from his European course of study.

He had also traveled to Europe to ask Grace Richards, the young woman who had helped Mary with chores around Wildwood, to marry him. After Grace had seen Merve and Denis through the early grief of Mary's death, she had returned to Europe to visit her family. Then Merve and Grace began writing to one another. The long distance correspondence by letters deepened their relationship. Grace accepted Merve's proposal. Returning to Wildwood via freighter in 1954, Grace found her own niche in the community working as a social worker in Nanaimo. She was good with foster children and often brought some of them home to spend the night at Wildwood. Both Merve and Grace fell in love with the first child, Michael, and eventually adopted him. Within a few years two more children were adopted at Wildwood—Marquita and Tisha. Denis graduated from high school, moved to Saskatchewan to work as a mechanic, and married. The marriage proved a challenge for Merve, though. "He has a fine wife and children, but because I am not of their faith, there is a division between us."

Logging Practices

With Merve's life getting re-established, it came time to make the third forest cut (1955–56). Merve spent hours walking through the woods taking measurements. Remembering his recent life tragedy, he did not want to take anything for granted. "I spent more time on cut number three than I've spent on any cut before or since. I wanted to make sure I got my forest back in balance by cutting the right proportion of each different species."

One of his challenges in caring for cut number three (and, in fact, all of his cuts) was finding honest buyers for his logs. Not only did he have to be an expert feller, a logging operator, and a forester, but he also had to be a businessman that held to his ideals. In 1956, right before his third cut, a buyer offered him a check for $56,000 to cut all his trees (forty times his original land purchase price). Grace was tempted, but Merve didn't bat an eye while turning it down. Nevertheless, the buyer was happy to negotiate with Merve for the logs he did cut because word had spread that Wildwood produced superior-quality wood. A number of factors contributed to that quality. Merve never cut over the annual growth (the interest) rate of the forest (the principal). He thinned for light to keep the canopy healthy and promote grain uniformity. And healthy mature trees were left for natural reseeding and to serve as standards for growth.

"I've cut some trees that are 1,400 to 1,600 years old," Merve says, launching into a familiar refrain, "but only because they were going to die. I do selective, sustainable logging. I have learned to never high-grade [take the biggest and the best] for any reason—because big, healthy trees produce the best seed. A forest only grows as tall as

---

European Forests

Europe has had to manage its forests by careful standards. Population densities are greater than those in North America, and harvesting of forests has been going on for thousands of years. The Black Forest in Germany is an example of a managed forest. It serves as the center of the country's timber and woodworking industry and is still a much sought after tourist spot. Each year specially selected plots of land are cleared of trees and replanted. Black Forest wood products are of high quality.

the most viable growing tree. The ancient Romans high-graded Lebanon and North Africa and turned the region into desert."

Merve's forest is like a community of diverse people. There are old, young, and middle-aged trees. Each serves a purpose to the collective. There are trees of different species—variety keeps the forest healthy if disease comes and destroys all members of one species. And as is true in both the forest community and the human community, soil is the basis for sustainability.

Merve recognizes that soil is his real resource, and he seeks to maintain the fragile balance of bacteria and other microorganisms, fungi, and nutrients that feed the forest. He minimizes the size and number of roads he uses for hauling. He is careful not to change the natural drainage patterns. And he understands that the most important thing he can do for the soil is to leave the proper amount of biodegradable waste

≥ *Merve's forest is like a community of diverse people. There are old, young, and middle-aged trees.*

Photo Courtesy: Ann Linnea

to rebuild. "At first I was a bit too tidy about the forest," explains Merve. "Slowly I learned to leave 5 percent, sometimes even 15 percent, of what I cut. Always, always in selective, sustainable logging, what is left behind is more important than what's taken."

Merve's objections to clear-cutting pale in comparison to his opinion about slash burning, a practice where after a clear-cut, debris (slash) is bulldozed into large piles and set on fire. Slash burning produces extreme high temperatures that dry and fry the soil. "The burning of slash is the worst possible crime perpetuated against forest soil. When a natural fire, such as one started by lightening, happens in a forest that has not been cut, the soil retains enough moisture to buffer the microorganisms. But in an area that has been stripped, the soil's balance is already in jeopardy, and slash burning is devastating."

Connections Abroad

In 1969, at age fifty-eight, Merve again traveled to Europe. He visited foresters from Holland, France, and Switzerland. (He had not yet appeared on the Cam Cathcart show. At this point he was still the student seeking further validation of what his forest was teaching him.) "I visited one sawmill in Switzerland that had been operational for 250 years," explains Merve. "They had, of course, modernized over the years, but it was still providing local employment because there was still wood to cut in the local area!"

Being in Europe reinforced Merve's belief that the product coming out of the forest must take second place. The welfare of the forest must come first. He returned to Wildwood with renewed enthusiasm for spreading the word about the importance of selective, sustainable logging. The lithe logger talked to lumberyards, sawyers, anyone who would listen to him about what he was doing at Wildwood. Because Yellowpoint was no longer isolated, now connected by paved roads, cars, and other conveniences to the rest of British Columbia, word of Wildwood began to spread. The man who dared to step beyond conventional North American forestry models was determined to help people understand that western forests were finite—that they needed protection and proper management.

Life handed the energetic forester other challenges beyond trying to educate people about the importance of shifting away from taking resources for granted to caring for resources. Grace asked for a divorce, and in the settlement Merve lost part of his land. His beloved parents died, and his son Denis remained estranged from him. But time, patience, openness to learning—all of the qualities that guided Merve in the tending of his forest—transformed these challenges into renewed personal strength and vision for the international prominence that was coming to Wildwood.

Merve's 1986 appearance on *Pacific Report* brought the world to his backyard. One of the first responses to the program came from Germany. A major university funded a graduate student to come and study alder regeneration for restoring their forest soils. Over the following years, professional foresters have come to Wildwood from Germany, the Netherlands, Finland, Norway, Sweden, Luxembourg, Switzerland, Spain, Jordan, France, Japan, China, Borneo, Java, Australia, Chile, Bolivia, Brazil, Costa Rica, Mexico, Cuba, Nigeria, Libya, and the United States.

Merve began receiving calls to speak at various gatherings. These offers connected the visionary woodlot owner with kindred spirits all over North America. One particularly memorable conference for Merve occurred in the town of McBride near the northern British Columbia–Alberta border. He listened to the story of this Mennonite settlement, which had decided it wanted no more clear-cutting and spraying for brush control. Though the government was willing to pay for helicopter spraying, the citizens said, "No more chemicals." Organizing every able-bodied person, they hand cut all the brush in a three-thousand-acre parcel for less money than it would have taken for one helicopter run.

Merve and the Raging Grannies

Merve came from activist stock. His grandfather Robert W. Wilkinson had been blacklisted from English mines because he advocated ten-hour working days and keeping children younger than age twelve out of the shafts and tunnels. "My parents were like my grandparents," says Merve. "Always left of center for the best causes." Following in their footsteps, Merve used his love for plain speaking and truth telling to give

⌃ *Professional foresters began coming from all over the world to the home of the man who was cutting trees to support himself and his family, yet he had more standing timber than when he began.*

him courage and stamina to be on the opposite side of the status quo. His first two wives helped weave him into the traditional roles of husband, father, and community member. But his third wife was the perfect companion for his elder activist years.

Anne Pask was a member of the Raging Grannies—a group that formed in 1987 to bring a spirit of humor and flamboyance to peace and environmental causes. When they appear at protests, the grannies wear flower-bedecked hats and shawls and engage in street theater and sing original songs. One of the premier interests of the Canadian Raging Grannies is ecoforestry, which, of course, led the group to tour Wildwood in 1989.

Merve was seventy-eight, and Anne was sixty-nine. Anne had never been married. Their courtship lasted two years, during which they supported each other in important activism. Anne testified against the British Columbia government for allowing nuclear-powered ships to dock in Victoria's Harbor. Some time later Merve testified before Great Britain's Prince Philip, then president of the World Wildlife Federation. Prince Philip had been flown to Canada's west coast by the MacMillan Bloedel Corporation to "have a look at the good job they were doing of preserving the natural habitat while clear-cutting the forest." Merve and thirty-one other delegates met with Prince Philip beforehand and cautioned him not to accept a "snow" job. As a result of Prince Philip's visit and ensuing report, Great Britain began to boycott British Columbia's wood and wood products unless they had come from sustainable forestry practices.

By 1991, Merve and Anne were ready to become husband and wife, and a large community wanted to celebrate the occasion. On April 21, three hundred people crowded into the First United Church in Victoria to celebrate their wedding. The Raging Grannies occupied the back rows of the church, and at the service's end, they formed an archway with their parasols for the couple to step out into the world. And step out they did. A year later Merve and Anne joined the protests over the clear-cut logging of western Vancouver Island's magnificent Clayoquot Sound.

Clayoquot Sound

Clayoquot Sound (pronounced clock-wit) is located in the center of Vancouver Island's west coast. Encompassing nine major watersheds, in the early 1990s it was still a pristine wilderness of lush temperate rain forest—a million acres of untouched coastal old growth the size of ten thousand Wildwoods. Clayoquot Sound was a treasure chest of giant cedars and hemlocks cascading down to fjords and across small islands that edged the coastline of the great Pacific Ocean. It comprised an area large enough to support a complete biosystem, including large carnivores such as mountain lions and bears. Here was the kind of forest backpackers tiptoed into. Here was the kind of forest that could heal the soul. And it was scheduled for logging.

When the government of British Columbia and the MacMillan Bloedel Corporation announced plans to clear-cut the sound, a cry rose up around the world. Local motel and restaurant owners, tour guide operators, and other business people pleaded with the government not to ruin an area that could be sold over and over for tourism. Citizens from eastern Canada journeyed west to protest the ruination of the last great forests of their country. Protestors came from other countries to speak to British Columbia government officials on behalf of this international jewel. In June 1992, protestors began assembling on the Kennedy River Bridge—the only road access into the remote forestlands.

Friends of Clayoquot Sound asked Merve to come and speak. As a forester, he knew "no logging" wasn't realistic. He convinced the protestors that "no clear-cutting" was a more reasonable goal. Protests and road blockages continued. Scuffles occurred between loggers

Photo Courtesy: Ann Linnea

and protestors. The Supreme Court of British Columbia issued an injunction against the blockade. On the morning of August 9, 1993, Merve and Anne joined 680 people at the bridge. All were arrested, charged with contempt of court, and bussed to the recreation hall in the town of Ucluelet for temporary holding. It was one of the largest acts of civil disobedience in Canadian history.

According to Merve, the trials of those arrested contained numerous irregularities, "not the least of which was trial by group, no jury, and no witnesses. Judge J. Bouck, the man who had issued the injunction, had been on the legal staff of MacMillan Bloedel before his Supreme Court appointment. Bouck was given the largest group for the first mass trial. The first thing out of his mouth when he walked into the courtroom was, 'I don't care what any of you say, you're all guilty.'"

"Can you imagine?" asks Merve.

Merve and Anne were tried with another group of protestors by Judge Skipp of the Supreme Court months later, early in 1994. "He was a decent enough fellow," says Merve. "At the end of our trial he actually told us we shouldn't be guilty, but because of the precedence of the other judges, we had to be." Anne and Merve received the stiffest sentence: one hundred hours of community service each. In his judgment statement, Skipp called Merve's testimony "magnificently unrepentant."

In May 2000, the Canadian prime minister, Jean Chrétien, designated nine hundred thousand acres as an International Biosphere Reserve. Under that designation no clear-cutting is allowed. A form of selective logging that uses helicopters to take a *few* prime trees in areas that are away from coastlines and stream sides is allowed. All logging is done by Iisaak, 51 percent owned by the Nuu-chah-nulth band, and 49 percent owned by Weyerhaeuser. (*Iisaak* is the Nuu-chah-nulth word for respect.)

The results of the Clayoquot protests are significant. "It took a social cataclysm in Clayoquot to shatter the old mold," reports Eric Schroff, general manager of Iisaak. Clayoquot was a huge step in the right direction, but in other parts of the province, tens of thousands of acres of prime old-growth British Columbia forest still remain unprotected. In 2001, Merve was inducted into the Order of Canada—the nation's highest honor.

The Future

Since 2001, Wildwood has been managed in partnership between Merve, the Land Conservancy of British Columbia, and the Ecoforestry Institute. Management and use of the land will be transferred to the Land Conservancy of British Columbia when Merve dies. As long as he is able, Merve will continue to live in the log house that he built so many years ago. Cluttered with piles of forestry articles, many containing references to him, the old house and its owner are full of stories.

Every day Merve continues to learn from his land. Walking with him among his trees is like walking with a druid. Firs, hemlocks, and cedars reach skyward, and the filtered light coming through the arboreal canopy illuminates an understory of many textures. Merve doesn't walk far these days. He pauses to rest on a stump he cut by hand fifty years ago. The stump and Merve have both become nurse logs now. The stump is supporting a young Douglas fir that has benefited from its slowly decomposing nutrients. Many young foresters have been nurtured in Merve's presence.

"There is a spirituality in these woods," Merve speaks from his resting place. "I can feel it—kind of a 'Mastermind' behind things. Whether that Mastermind is a He or She doesn't matter. What matters, I believe, is that we're custodians of the earth while we're here."

Few people have cared for one piece of land as long or as carefully as Merve. He has performed his custodial duties well. The Mastermind of Wildwood is undoubtedly pleased. When this old one falls, the reverberations will be felt in forests all over Vancouver Island and beyond.

Postscript

- Anne lived with Merve at Wildwood until her death in 2004.
- At the time of this book's printing, Merve was alive at age ninety-eight and still living in Wildwood.
- Wildwood will remain intact, preserved, and selectively logged under the management of the Land Conservancy of British Columbia.
- The eagle still lands in the old cedar, and the woodland birds still sing.

When this old one falls, the reverberations will be felt in forests all over Vancouver Island and beyond. »

Photo Courtesy: Marcia Wiley

Photo Courtesy: Ann Linnea

Name: Laura Robin

Occupation: Pruner, builder, masseuse, chef, and writer

Point of Wisdom: Pruning trees is an art form that can best be learned through apprenticeship with the trees themselves.

CHAPTER 3

The Tree Pruner

Laura Robin

I prune according to what I see and
hear what the tree is telling me.

—LAURA ROBIN

LAURA ROBIN'S TREE KNOWLEDGE comes from thirty years of spending time climbing, pruning, and studying them. Laura, now in her sixties, has a lean, weathered frame that resembles the trunk of a native Montana pine. Her insights about life and pruning are as basic and solid as the trees she loves and cares for.

Laura also loves a little cabin perched in the foothills above Bozeman, Montana. Most winters the dirt road to the cabin is not plowed, so she skis the last quarter mile to her hideaway. In spring, summer, and fall, it's less than an hour's drive from downtown. Nestled into a hillside of Douglas fir, Ponderosa pine, and lodge pole pine, the cabin is Laura's heart home—a place of refuge. She learned to dance watching Montana winds sweep in from the plains and twirl those trees. And when she recently remodeled the cabin, she enlarged the front deck so she would have more room to dance with the wind.

As a child, she rode her horse through the forests and meadows of this family land near Bear Canyon. She found young willows to make fishing poles and whistles. And she helped her father build a log barn. "Trees were always a part of my world as a kid," she explains. "There was no separation between me and nature. No way could I imagine a world without trees."

"Even when I was tiny, I remember loving trees," says Laura, pulling in another story with the ease that a tree catches wind. "Mom always made me take naps. I remember being about five or six and pleading not to take naps. Our compromise was

The wild scenic beauty of Bozeman, Montana, birthed Laura's life. ⌄

that I got to take them under the green ash tree in the backyard of our house in town. I loved that tree. I felt safe lying next to it."

As trees grow from saplings to adults, they encounter challenges that literally shape their survival—lightning strikes, prevailing winds, and competition for light and water resources. Life presented challenges that shaped Laura, too. "There was a moment that I can remember like it was yesterday. I was sitting in a classroom at Montana State looking out at this incredible diffuse light coming through the campus trees. I didn't want to be in class; I wanted to be out in nature, the elements, the reality of life. It was like a neon sign telling me to get the hell out of town and learn something about life. So, I walked away from everything I knew and headed to California because that's where, in the late '60s, everything seemed to be happening." And then the storms began.

It was 1967. The streets of California were alive with protests against the Vietnam War, experimentation with drugs, and rising demands for women's liberation. California was a siren's call for a country girl from Montana to go test her wings. Laura's aunt lived in Southern California, so she went to live with her and landed a government job with top security clearance. Within months Laura was accused of being a lesbian and fired. She drifted north to San Francisco and drifted deep into a subculture she was not prepared to handle.

"I pretty much lived on the street where I met people who were muggers, prostitutes, and drug addicts. I didn't care who they were because they accepted me," says Laura, "but I ended up in some scary situations—getting beat up, waking up in houses of people I didn't know. On some of those dark nights, what saved my life was remembering the sunrise coming over the trees of Bear Canyon . . . remembering that there were other places I had come from and other ways to live."

Trees grow by following a leader—a vertical branch or extension of the main trunk—that seeks height, light, and core stability for the weight of the tree concentrated through the trunk. A leaderless tree, where this structure has been broken off or removed, is more vulnerable to disease and decay. A broken leader is a fairly common occurrence in the weather a tree endures in the course of a lifetime. In response the tree will usually sprout a number of small branches around the point of injury, and a period of confusion follows. For the tree to grow true again, one branch needs to assume dominance, go vertical, and re-establish the tree's core stability.

Laura essentially lived on the streets for the next five years and almost lost herself in the leaderless experimentation of the times. "Slowly I started getting on my feet again," she explains. "One of the people who was kind to me then was an Italian

The idea that she could find a profession that would keep her outdoors germinated in the dark soil of her true self. »

Photo Courtesy: Ann Linnea

man who ran a little dry goods store in North Beach. I was working for Bell Telephone and renting a room from him. We'd share a shot of whiskey under the counter now and then, and we talked a lot. One day after listening to my mountain stories, he told me I should become a tree pruner. I don't know how he came up with the idea; maybe he just intuited that I belonged to the earth, and pruning was as wild a craft as he could imagine from his city experience."

Laura kept the phone company job for another year, but the idea that she could find a profession that would keep her outdoors germinated in the dark soil of her true self. She began pruning away competing interests and felt her connection to the natural world slowly returning her core strength. The tree woman was about to regain a dominant leader.

Pruning in Vancouver

"About a year later [1972], a friend and I drove up to Langley, British Columbia, near Vancouver. Fate or something introduced me to a woman there named Sharon who offered to teach me to prune. I had absolutely no money or transportation. Together we picked out a purple 450 Honda motorcycle. She gave me one handsaw, and we started working side by side trimming trees, driving from one job to the next on that motorcycle."

It was in this urban environment that Laura would practice her pruning skills. In a wild forest, pruning occurs naturally. Storms come and blow off branches or topple one member of a stand, and its fall takes down neighboring trees and sheers limbs off a half dozen surrounding trees. Deer, elk, moose, rabbits, and other browsers wander by and graze on low-lying branches or bite the leader off a twelve-inch pine. And trees prune themselves.

The trunk of a tree is like an accountant: it registers the amount of nutrients sent up from the roots and the amount of nutrients sent back by the branches. When a branch is no longer producing as much as it is taking, the trunk responds by "sealing off" this branch. When the first strong winds of autumn arrive and toss the trees around, many of these branches are torn free. They fall to the forest floor and eventually decompose—becoming part of the cycle of life, death, and regeneration.

In cities and towns, people prune trees for a variety of reasons—to shape them for beauty, to improve fruit production, to trim away hazardous branches, or to minimize blocked views or entanglement with electrical wires. A popular misconception is that anyone with a saw can prune. Unfortunately, more trees are killed or ruined each year from improper pruning than from pest damage.

The process of learning a new trade is often slow, not unlike the tree's process of laying down one layer of cambium after another—carefully marking its growth rings. Laura was a cautious, determined student. "Sometimes I'd start in on a little branch that didn't look like a big deal to me, and Sharon would call out, 'Stop, not that one! The tree is raising that branch up to be a leader, don't take it out.' Sharon didn't know it, but I cried in a lot of trees. I didn't want to hurt them. I felt the responsibility of decision-making, the pain of cutting, as if it were my own flesh and blood. Sharon forgave my mistakes, and I started finding a place in myself—a place I knew was genuinely me. Caring for those trees was the path back to caring for myself."

Laura was learning to prune, but it wasn't easy to start a new profession in a new place. She and Sharon often felt they had to prove that they knew what they were doing because they were women. One of the lucky breaks Laura got was the chance to work with an eighty-seven-year-old nearly blind orchard man named Tom. His father had been a master gardener for the Queen of England. "Tom had cleared land in White Rock and planted every kind of fruit and nut tree you could imagine—apple, plum, pear, walnut. A friend of his saw my newspaper ad and told him about me. He hired me to prune because he was losing his eyesight. He stood right beside me and directed every single snip of the pruning shears. I learned by practice, practice, and more practice. He picked up my education where Sharon left off and taught me the subtlety of small branches."

Trees as Teachers

The International Society of Arboriculture recommends consideration of the following principles before pruning a tree:

- Each cut has the potential to change the growth of the tree. Always have a purpose in mind before making a cut.
- Proper technique is essential. Poor pruning can cause damage that lasts for the life of the tree. Learn where and how to make the cuts before picking up the pruning shears.
- Trees do not heal the way people do. When a tree is wounded, it must grow over and compartmentalize the wound. As a result, the wound is contained within the tree forever.
- Small cuts do less damage to the tree than large cuts. For that reason, proper pruning (training) of young trees is critical. Waiting to prune a tree until it is mature can create the need for large cuts that the tree cannot easily close.

Laura had been mentored in these principles, and as she gained confidence, she began pruning on her own. She began to trust the trees to teach her what she needed to learn. "I'll never forget climbing into one particular apple tree. It was old, tough, and totally tangled—suckers shooting out every which way from the main branches. I was fifteen feet off the ground, and it started slapping me around. I mean I really couldn't move, and I began to feel unsafe." Having never experienced such resistance before, Laura climbed down from the tree, walked a complete circle around it, and said out loud, "What the hell is the matter with you?" She stood about ten feet from the trunk of the tree, arms folded, fuming mad. The longer she stared at the tree, though, the more things she began to notice.

"I started to really see what was going on with the tree. It was such a rat's nest. It hadn't been tended in years. As I grew more and more quiet, I noticed huge limbs had been sawed off. After a while, I got an image in my mind of a man having a fight with his wife and taking out his anger by slaughtering the natural symmetry of the tree. From that moment on, I knew I'd try never to cut a tree or trim a hedge without first being quiet and listening," Laura says softly.

"Pruning is tender work. When I'm pruning, I'm like a surgeon. I know I'm cutting into flesh and blood and what I do can cause pain or relieve pain. The apple tree and I had to be in harmony before I was safe in its limbs and before the tree was safe in my hands." Pruning is a way to bring out the tree's story. Just as a person uses a story to organize life into a pattern she or he can live with, so good pruning establishes a pattern of growth that is healthy for a tree.

From the story of the apple tree, Laura moved quickly to talking about a magnolia tree that her friend Yolanda had owned. "It was one of the most magnificent magnolia trees in all of Vancouver, which is no small thing in a city of gorgeous

Vancouver, B.C., is a city of beautiful trees. It was a good place for Laura to learn proper pruning techniques. ⌄

Photo Courtesy: Ann Linnea

trees. I pruned it every year for several years. One year I told my friend that her tree needed a bath because it was having a hard time breathing through all of the accumulated scale and moss."

Using carefully prepared insecticide soap and a brush, Laura scrubbed moss off the tree's main branches and then hosed the entire tree down. She was thrilled to work for someone who cared enough to pay her to give a tree a bath. "It was late afternoon by the time I finished. Yolanda came out, and we stood together, speechless and awed by the golden energy given off by that tree."

Laura's story suddenly stops. She chokes back the tears and then blurts out, "The next people who moved in that house cut the tree down. They understood nothing! The tree was the centerpiece of that front yard, the queen of the corner. To this day neither Yolanda nor her husband can stand to drive by their old address."

From Vancouver to Edmonton

Like so many people walking the edge between poverty and sustenance, Laura did a lot of things to pay the bills. For a time she even worked as a chef, but always she kept pruning. "Nowadays you need a piece of paper to prove you have knowledge about something. But I mostly have come to things by experience—not that I don't take professionalism seriously. I even made a portfolio of references to show people I was a genuine pruner. But you know something, most people don't care enough about their trees to ask for references."

After a few years, Laura's tutelage from trees, tree owners, and other pruners took her to Edmonton, Alberta, where she worked for a pruner and nurseryman of forty years experience. "I learned a lot from Ed. Pruning hedges and shrubs is a whole different ball game than fruit trees. He encouraged me to climb ladders higher than I ever had. He taught me about the dangers of spraying, about how to remove a stump with rock salt, and about not topping trees. He was a good man."

Working for Ed, Laura had the benefit of a regular paycheck, exposure to a wider variety of pruning opportunities, and a chance to work with larger crews. "I often found myself one of the oldest members of an all-male crew. However, I generally found I could climb higher than any of them."

Laura's high standard for the quality of her work has always been based on her loyalty to the trees, more than loyalty to her clients. Nevertheless, it was important to her to be well paid for what she did. "One day, maybe three or four months after

After years of absence, Laura found her way home to Bozeman and to Bear Canyon. »

Photo Courtesy: Stormy Apgar, Thunder Cloud Images

I had been working for Ed, a man whose hedge I was trimming told me how much he was paying Ed for *my* work. It was a hell'va lot more than Ed was paying me. So, I decided right then and there to go into business for myself."

By this time in her pruning career, Laura had upgraded from traveling to and from jobs on a motorcycle to using a van that carried a ladder, an extension pruner, a handsaw, loppers, and hand pruners. One of the things she had to learn about being in business for herself in a new area was costing out jobs. Launching on another story, she says, "It was a cherry tree that taught me how to estimate."

"Cherry tree root systems are humongous compared to what's above ground," Laura gestures with arms wide apart. "I bid on cutting down and digging out a cherry tree in a guy's alley. I swear it nearly killed me. As I got into the job and realized how much work it was really going to be, I went back to the man and tried to renegotiate. He actually got mad and said I had to do the job for what I had estimated."

"So, load by load, I took away all the debris I had so carefully cut. Finally all that was left was the root ball lying in the alley where the tree had formerly stood. I had no idea how I was going to move that heavy sucker. So, I slept in my van overnight hoping I'd figure it out."

"When I woke up the next morning, there was a young man walking down the alley," Laura says. "He came right over to the van and announced, 'I've been wondering what it would take to move that thing.'"

"You're my man," Laura smiles in recall.

"We used a crowbar and lots of pushing and shoving up an old wooden plank. Somehow we got that root ball into the back of the van. Geez, he was a nice guy. I didn't make a dime on that job, and I doubled my prices right on the spot."

Tree pruning is a hazardous profession. Thousands of homeowners are injured or killed every year attempting to prune their own trees. Professional pruners or arborists must now pass rigorous certification requirements to ensure that they attend to proper safety and pruning standards. Personal protective gear is basic for those who prune: goggles, hard hat, long sleeves and pants, and good boots. Pruners need to take great care in the choice and placement of ladders. Tools must be sharp and handled carefully. Attention to weather for added dangers due to wind and rain is crucial.

Like many pruners, Laura uses both ladders and tree climbing to get her work done. She trusts the trees. "A tree won't throw you," she says emphatically. "I have never had a tree drop me. Once on a rainy day I slipped and fell nearly twenty-two feet in a silver maple, but she caught me twenty feet from the ground. I had a few bruises, but no broken bones."

Like most pruners, though, she did suffer one serious fall. Eight years ago she fell off a ladder while pruning a twenty-foot laurel hedge. She broke her elbow, cracked her pelvis in two places, and severely traumatized her right knee and ankle. In true survivor style, one week out of the hospital she started pruning again—left-handed, standing on the ground. "The first time I stepped onto the second step of a ladder, I broke into a cold sweat," she remembers. "But I knew I had to get out there and work again. I was fifty years old. I was learning my body had limitations. I couldn't see it then, but more change was on the way."

Laura fully recovered from her pruning accident. ⩒

Photo Courtesy: Ann Limnea

Home to Montana

In March 2000, Laura came home to Bozeman when her mother was diagnosed with terminal cancer. For the better part of a year, she devoted her life to caring for her mom. In that year, and in the years since, Laura has reconnected with the land and trees of her child-hood. "I now know our Bear Canyon land like nobody else has ever known it. I've walked every foot of it. Sometimes I've crawled on it, looking for help to hold all of the grief I carried home. People I knew decades ago and people I've just met have been good to me, but sometimes the only thing large enough to hold a life's worth of grief is the earth herself."

Photo Courtesy: Ann Linnea

One of the first things Laura did after her mother died was to shore up the one-room cabin located on her beloved land. At the edge of the meadow on the other side of the hill from the cabin, she created a simple prayer bench out of stacked stones and an old piece of lumber. From this humble perch she can gaze at Bozeman far below or study the movement of elk, deer, cougars, and bears across the dry foothills. And, of course, she has a special tree.

"My tree has been hit by lightning, just like I feel I was hit by the loss of my dad and mom. The tree hung on to its roots—just as I am now hanging on to my roots. At this elevation in this dry climate, Doug firs don't get as big as they do on the West Coast, so the tree is small, yet still quite old. I call her my journey tree. With her help I can call myself into really open communication with the spirits of nature in all forms. I feel connection and belonging." Standing next to her journey tree, Laura mirrors its

⌃ *Rebuilding the cabin was the rebuilding of Laura.*

hard-earned lines of weathering, deep-rootedness to the earth, and strong life force. They are both beautiful.

Since her mother's death, Laura has been sorting through what to do with her life. She graduated from massage school and still prunes the trees of people who love their trees. "I charge them reasonable money, and I love up their trees. And I also know that when I prune all day, I am totally exhausted. There is no creative energy left in me. So, I'm pretty selective about the jobs I take." And Laura has kept working on the cabin up at Bear Canyon. "Rebuilding that cabin has been the rebuilding of me," she says. Laura installed a composting toilet, built kitchen cupboards out of bamboo, laid a fir floor, and put knotty pine in the ceiling. Using five cords of wood to heat each year, Laura has surrounded herself inside and out with the comfort of trees.

The woman who was loved and protected by trees in her youth, who was pruned by the sharp edges of street life, has walked the long path back to communion with the trees that first held her. She has restored her internal leader. The remaining decades of her life will undoubtedly be about reaching the full strength of standing in place.

"Oh Healer" (October 23, 2001)
Laura Robin

You are a hundred feet tall
in my eyes
I know you have picked me
to dance for
How free you are with feet
so firmly rooted.
Hold me in your arms
and let the winds flow through
my mind.
I hear your blood quieting
with the coming of winter.
I need to stand on your feet
so deep in the earth
and join you in the grounded
Of grounding center.

Photo Courtesy: Ann Linnea »

Photo Courtesy: Ann Linnea

Name: Bud Pearson

Occupation: Wood turner

Point of Interest: Working with wood can be a source of tremendous healing and education. The artist must understand his medium and all the environmental ramifications for its use.

CHAPTER 4

The Man Who Turns Wood

Bud Pearson

Being a turner, I look at trees differently. Many people take them for granted. They maybe notice that trees are large or small. But because we understand wood's subtle differences, wood turners recognize that every tree in nature has a purpose.

—BUD PEARSON

THE TINY WOODEN BOWL IS LESS than an inch in diameter and about the weight of a hummingbird's egg. Bud Pearson has crafted it from buffalo berry, a shrub that grows on the hillsides around Bozeman, Montana. Bud's powerful hands gently cradle the cinnamon-colored thimble, and though his tanned, weathered face gives him the appearance of a cowboy or landscaper, he is a retired postman and an artist.

carver. The carver works with handheld knife and chisel and mallet to shape a stationary piece of wood. The turner presses a variety of metal tools called gouges (scrapers, skews, and other specialty tools) against the rotating piece of wood. Like a potter's fingers on a lump of spinning clay, the gouges shape the spinning wood.

As Bud pushes his carefully chosen tools against the wood chunk, chips fly in every direction. His experienced hands gauge how hard to push, what angle to move, when to back off. He works a tiny piece of birch into the shape of a pen body. The dance between his hands and the turning wood is an intricate ballet, which both releases the sweet smell of a birch tree in the August sun and coats the floor with sawdust.

Learning the Art

While Bud makes micro adjustments to the pen emerging on the lathe, he explains, "Some of the best turners are Europeans. They come up through the guild system

≳ *The wood for this birch pen was cut several years ago and has been drying since. A turner must move at the pace of his medium.*

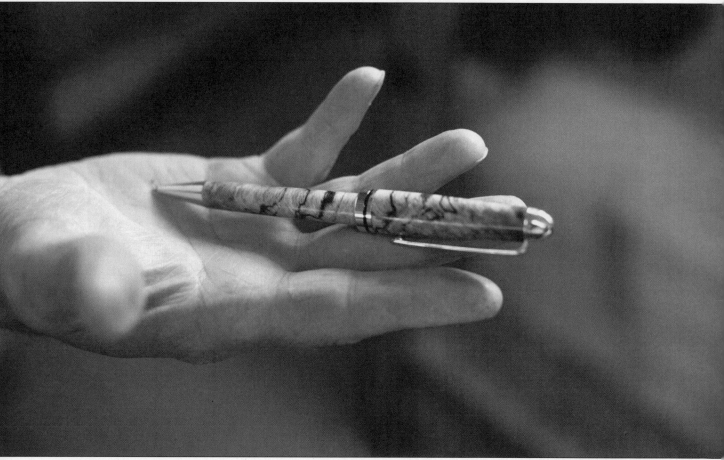

Photo Courtesy: Ann Linnea

The Man Who Turns Wood

Bud Pearson

Being a turner, I look at trees differently. Many people take them for granted. They maybe notice that trees are large or small. But because we understand wood's subtle differences, wood turners recognize that every tree in nature has a purpose.

—BUD PEARSON

THE TINY WOODEN BOWL IS LESS than an inch in diameter and about the weight of a hummingbird's egg. Bud Pearson has crafted it from buffalo berry, a shrub that grows on the hillsides around Bozeman, Montana. Bud's powerful hands gently cradle the cinnamon-colored thimble, and though his tanned, weathered face gives him the appearance of a cowboy or landscaper, he is a retired postman and an artist.

"There's other things I do, but wood turning defines who I am," he explains. "The pieces I've turned give me a feeling of joy and satisfaction. Even this little bowl here, well, I get the sense that I've done something worthwhile." His smile lights his face like the rays of wood dust that catch the sun streaming into his shop.

Although he has since moved to Mountain Home, Idaho, Bud got his start as an artist in a small, wood frame house in an older Bozeman neighborhood, a few blocks off Main Street. The homes there have small, carefully mowed lawns with one or two large maple or ash trees set in the middle for summer shade and winter wind break. These are urban trees bordered by flowers, lilac bushes, and the horizontal reach of junipers. His old shop was an eight-foot by ten-foot room built on the back of his house, hidden from view. His new shop is in his double-car garage. People passing by in either neighborhood would have little clue that these modest homes house an artist. Actually, Bud is shy about the title "artist." As he says, "A lot of turners are craftsmen, few are artists."

When Bud dons his canvas helmet with air filtration hose (for eye and lung safety), this small, graying, midlife man is transformed into a wood turner. Each piece of wood he turns has gone through a prolonged drying process that begins with air-drying in the yard and ends with drying time in his laundry room. Some pieces dry for months; others dry for years. It makes for interesting décor: wood-paneled walls, a small wooden kitchen table, and chunks of wood quietly dehydrating, like cats who never move. Bud lives with wood. It is his quiet roommate. Around the house, often stacked in precarious piles, Bud's bowls range in diameter from seven-eighths of an inch to twenty-one inches. They are crafted from maple, oak, cherry, plum, and willow or from more unusual woods like the shrubby buffalo berry and sagebrush. He also turns burial urns, Christmas tree ornaments, wooden balls, and pens. "You have to turn for grins and giggles, too," he explains, displaying a whimsical ornament.

This afternoon, his voice slightly muffled in the canvas hood, Bud explains the process as he sets about clamping a piece of carefully prepared wood securely onto his lathe. A lathe is a horizontal machine that holds the wood firmly in place and spins it at a consistent speed. When Bud flips the electric switch, the lathe rotates the attached chunk of wood, and he sets about shaping the piece by applying pressure with varying sizes and shapes of sharp metal tools. It takes skill to apply the tools with the correct pressure to elicit the shape from the wood.

After watching the wood chips fly, one can imagine the craft it might take to turn a sturdy salad bowl; it's harder to imagine the art required to create Bud's delicate hummingbird bowl. The use of a lathe distinguishes the wood turner from the wood

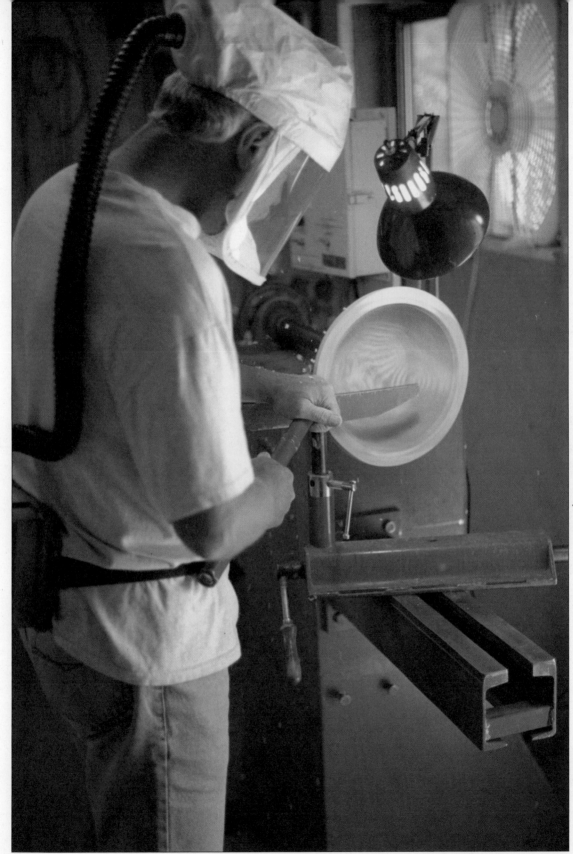

⌃ *Dressed in safety gear, Bud demonstrates the use of a lathe for working a piece of wood into one of his fine bowls.*

carver. The carver works with handheld knife and chisel and mallet to shape a stationary piece of wood. The turner presses a variety of metal tools called gouges (scrapers, skews, and other specialty tools) against the rotating piece of wood. Like a potter's fingers on a lump of spinning clay, the gouges shape the spinning wood.

As Bud pushes his carefully chosen tools against the wood chunk, chips fly in every direction. His experienced hands gauge how hard to push, what angle to move, when to back off. He works a tiny piece of birch into the shape of a pen body. The dance between his hands and the turning wood is an intricate ballet, which both releases the sweet smell of a birch tree in the August sun and coats the floor with sawdust.

Learning the Art

While Bud makes micro adjustments to the pen emerging on the lathe, he explains, "Some of the best turners are Europeans. They come up through the guild system

≽ *The wood for this birch pen was cut several years ago and has been drying since. A turner must move at the pace of his medium.*

Photo Courtesy: Ann Linnea

and have tremendous training and a sense of tradition. In America, we are informal learners. Turners here learn from one another at symposiums and classes and through personal exchanges. We love to stand around and talk wood. It's a hobby for a lot of retired men who love to work with their hands."

After serving in the Vietnam War, Bud returned to his hometown, Bozeman, and became a letter carrier. The daily routine of walking and the stability and familiarity of the route helped this man of few words integrate the harsh wartime experiences that haunted him. Letter carriers often get to know their customers. One of the folks on Bud's route was a wood turner named Steve Gray who turned high-quality, beautiful kaleidoscopes. Not far from his home lived a bowl turner named Gordon McMullen. Over time, at Gordon's and Steve's persistent invitations, Bud began stopping by after work to learn more about wood turning. Eventually Gordon built Bud's large lathe for him in 1992.

Working with his hands was as therapeutic for Bud as walking. "I grew up on a dry land grain farm with a father who wasn't afraid to fix anything. But as I got into turning, I realized that I enjoyed making things more than I enjoyed fixing them. Steve and Gordon were fabulous, genuinely eager to share their knowledge and skill.

"Gradually I met other turners, and I learned that many of the older ones had grown up making things with their hands. They had learned patience and repetition from the time they were young, something youngsters today in their digital lives often don't understand. The biggest challenge for people who want to work with wood is the patience required. We live in a fast-paced world. People expect to do things fast. Wood doesn't allow you to move fast. Wood slows you down."

A turner must move at the pace of his medium. Even Montana's and Idaho's fast-growing trees, such as willows, require decades to lay down enough growth rings to create a trunk that might become a bowl or ornament. And stronger, slower growing wood, such as cedar and fir, require a turner to spend years of practice to acquire the skill to work with them.

"The wood for this little pen was cut several years ago and has been drying ever since. It had to cure. I had to set it aside and wait. Sometimes people bring me a unique chunk of wood they think I'll like. I thank them and let them know it might be ten years before I can use it. That always takes them by surprise."

Bud knows that being in a hurry will only cause problems. "Usually when something is going wrong on a piece, it's a human problem, not a wood problem. Turning should create a sense of connection between me and the wood," and here he pauses and smiles, "But I have to say that some of my best pieces are those I've had a war with."

War. The word jars the sunlit workshop, serves as a reminder that Bud spent time in Vietnam, and like many vets, there's a part of him still spending time in Vietnam. Decades of walking and wood turning have rounded his edges, have softened his language, but "the war" remains a point of reference for him—the years his life was on the lathe.

The Mark of Vietnam

Just as the passage of time has healed Bud's war wounds, time has also healed parts of Vietnam. *Healed* may not be the right word—for soldiers, for populations of citizens, or for the ground on which battles have raged. But it is the hoped-for goal. Today, sections of the Vietnamese countryside have been refashioned as tourist destinations. Advertised as a place of "sublime beauty" with "divine beaches," "soaring mountains," and "dense, misty forests," this tiny country is attracting an increasing number of tourists from Europe and North America. Vietnam today provides an experience of great contrast: bikini-clad young women on hotel beaches, a museum to the country's liberation, tours of the demilitarized zone (DMZ) with aging soldiers posing on rusted artillery, and mile-wide strips of struggling scrub, the result of decimation by aerial defoliation, next to patches of tropical rain forest wonder.

Since the early 1990s, Vietnam has also become a destination of high interest to biologists. In this period of increased political stability, biologists have been allowed to explore remote areas such as the forested Truong Son mountain range on the border between Vietnam and Laos. Although local people have long appreciated the tremendous biological diversity of the area, biologists have been amazed to discover both large and medium-sized mammals previously unknown to science.

One newly discovered species is the saola, the only member of *Pseudoryx*, a genus new to the cattle family. At a weight of 220 pounds, the saola is the largest land-dwelling mammal "discovered" by science since the kouprey (gray ox) was classified in 1937. Also recently classified are three previously unrecognized species of barking deer, one species of pig, and one species of rabbit—to say nothing of three new species of birds, nineteen species of amphibians, sixteen species of reptiles, tewnty nine species of fish, and 516 species of invertebrates. All of this in only one small region of Vietnam!

Just as the end of war opened up the country and the man to new discoveries, it has also left its mark on both the country and the man. Visitors in the former DMZ (the boundary between North Vietnam and South Vietnam) are guided through the scorched land still in shock from the war. In the country as a whole, 60 percent of the tropical rain forest was destroyed during the war. (If that war had been fought on U.S. soil, this would be the equivalent of destroying all forests west of the Mississippi River.) In the complex ecology of forested lands everywhere, and most especially in the tropics of Southeast Asia, a few decades are not nearly long enough for nature to restore the complex system that had developed there over thousands of years.

Though Bud has been able to create a sane, productive, and enduring life out of the ravages of war, the restoration of one man's body and soul is also a complex process. Healing the heart, like curing the wood, takes its own mysterious time. Not surprisingly, as a man of the earth, Bud still has strong feelings about the effects of the war on the earth of Vietnam and about the ethics of combating armies in how they treat the land on which they fight.

"Several writers in the Old Testament specifically mention that invading armies should not cut down the fruit trees of their enemies. It was considered immoral to steal the future from people, even in war," says Bud. The specific instructional passage from the Bible Bud is referring to comes from Deuteronomy 20:10: "When thou shalt besiege a city a long time, in making war against it to take it, thou shalt not destroy the trees thereof by forcing an axe against them: for thou mayest eat of them, and thou shalt not cut them down (for the tree of the field is man's life) to employ them in the siege."

"Of course, this ethic has been overridden many times, and whenever that happens, humanity takes a deliberate step into barbarity," continues Bud. "When the U.S. and its allies went into Cambodia, entire mahogany plantations were destroyed because they were cash crops. And in both Cambodia and Vietnam, Agent Orange and napalm were used to devastate vast sections of the natural jungle, including huge ancient trees, so the enemy would have no place to hide. We also let these chemicals drift onto adjoining rice lands and into people who ingested these poisons through their food. We called it Herbicidal Warfare. We knew what we were doing to the land. The old-growth forests of Cambodia and Vietnam were priceless global treasures that are gone, perhaps forever."

It's a long speech for a quiet man. A listening silence fills the shop. One senses Bud has spent years painfully putting together these thoughts. Between March 1965 and November 1968, North Vietnam was bombed by "Operation Rolling Thunder"— more tonnage of bombs were dropped on this tiny country than were used during the entire Pacific Theater of World War II. Two million acres of land was destroyed, and an estimated six million bombs and mines remain embedded in Vietnamese soil. One of the healing gestures toward these ravaged lands was launched in 1995 in Seattle, Washington. PeaceTrees Vietnam was established to reverse the consequences of this war. Focusing its efforts on Quang Tri Province, just south of the DMZ, PeaceTrees Vietnam is slowly clearing land mines and then helping residents to re-establish agriculture and plant trees.

Just as wars profoundly change nature, they profoundly change the lives of people for generations. As a foot soldier in Vietnam, Bud strove to survive, just as the young men in the opposing army were doing. For a time after the war, Bud did what the post office told him to do. Now, however, Bud does what the wood tells him to do.

"Quitting the post office gave me a little pension to live on. I realized I had spent very few holidays with my parents or my brothers and sister. I can do that now that I have my freedom. Being closer to family is part of the reason I moved to Idaho. And becoming a turner has given me a chance to bring things of beauty into the world."

Understanding Wood Variety

The pen taking shape on the small tabletop lathe is not anything like a mass-produced plastic or metal pen from the drugstore. "I'm turning birch that has been spalted—that means it has a dark mold that has grown through into the grain. Mold makes an interesting pattern, which is a way turners work with whatever has happened in the life of the tree. Before I started, I covered this piece with superglue to make sure it was hard enough to handle being turned. Next in the sanding phase, I have to slow the lathe, or the wood could heat up and burn or check (crack).

"Even this one little piece of wood has a lot of lessons for me. It might be telling me to slow down or to sharpen my tool. I have to pay attention. There is a certain feeling of joy and adventure that comes with each piece. Otherwise, I might as well be on an assembly line."

Bud has spent years honing woodworking skills and his very private philosophy about wood turning: "The attitude in this country, that we can go to a big box store to buy anything we break, instead of fixing it, is putting us in a world of hurt. I hope in some small way my work reverses that trend. By making something well so that people *want* to care for it, I hope they pass along my pieces so that their grandchildren will also care for and use them."

As he turns off the lathe and lays his safety helmet on a sawdust-covered metal bench, Bud keeps talking. It's a warm day, so sunlight and summer breeze filter through the open door, enhancing the meager light from one overhead fixture and two small

Photo Courtesy: Ann Linnea

windows. Bud moves from his shop into his house to the laundry room where shelves are lined with many pieces of drying wood.

He starts pulling out various chunks he is drying for future use and explains the varieties he keeps around: "These pieces have come from the tropics. Tropical woods are dense, heavy, and oily because they have to resist bugs. They also contain toxins, which make them unusable for food bowls. These pieces here are local willow and orchard woods like cherry, apple, and plum. They have to dry a really long time— sometimes years—otherwise when the gouge hits the turning wood there's literally water flying everywhere in the shop. Once in a while, though, I do pre-shape green wood before letting it dry. And these slabs of old-growth cedar, fir, and hemlock are from windfalls in regional forests. They're always in high demand because the grain is tight, making it very strong."

Bud doesn't deny his own impact on trees. "The second year after I retired and was turning almost every day, I took over 5,600 pounds of wood chips to the dump. That came as a big shock! Woodworking brings me right into environmental issues—clean air, clean water, preserved forests, and correct land usage. Wood turners know that if we take a tree out, we should plant one back. And we know that not using something is equally wasteful. If a tree falls down from weather or is taken down by foresters, there are always parts of it no one else wants that a wood turner can use. We like interesting chunks."

Like many wood turners and carvers, Bud does not work with tropical woods, unless, like the small chunks in his shop, they have been given to him. Norman Myers, a British environmentalist and authority on biodiversity, calls tropical forests "the most exuberant celebration of nature that has ever graced the face of the planet." Myers goes on to explain that although they occupy only 5 percent of the earth's land surface, tropical forests contain over half of all species on earth. While biologists are actively exploring intact rain forests, such as in Vietnam, they are racing against the chain saw and other development. More than half the world's rain forests have been lost, and they are disappearing faster than any other ecological zone.

Rain forests are dwindling because every issue in the world is tied to every other issue in the world. Consumer demands in North America and Europe and now also China, India, and Russia are irreversibly impacting global ecology. Developed countries assume we need more oil and if we can't have it, then we'll just substitute some other product, adapt our engines, and keep going. So-named developing countries, like those containing the Amazon basin and the rain forests of southeastern Asia, have land that holds oil reserves or land capable of growing either soybeans or palm oil for biofuels. Megacorporations come in promising jobs and a piece of the action. The forestlands are cut, drilled, or planted. This cycle repeats the problems caused by developed countries demanding rubber, bananas, or coffee.

In addition to the problem of lost species (many yet undiscovered), cutting these forests releases carbon into the atmosphere and contributes to global warming. According to a 2007 Global Canopy Program report, tropical deforestation contributes nearly one-quarter of global carbon emissions, second only to the burning of fossil fuels.

≽ *This Saturn bowl is made from oak and is so-named because it wobbles on the bottom.*

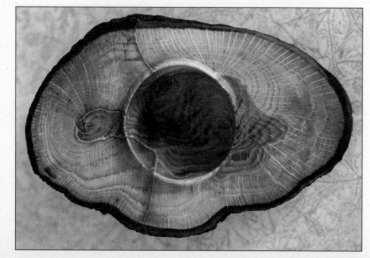

Photo Courtesy: Ann Linnea

Part of the way Bud handles his environmental commitment is to harvest and gather wood locally. "Eighty percent of the wood I use I gather around here, much of it from city forests. I try to get it in winter when most of the moisture is in the roots. Around Bozeman and Mountain Home, I talk to city foresters whose job requires taking trees down for new development or pruning around power lines or cleaning up after a storm. Since I'm a small-object turner and the amount of wood I need is minor, they are usually glad to oblige."

Bud has traditionally been ambivalent about selling his work. Although he has worked on commission with a number of local galleries, he often gives away his pieces. When he first moved to Idaho in 2007, he was trimming a tree and managed to break one arm and strain the other. That put a temporary halt to his turning, but not for long.

This gentle, thoughtful man is like a piece of exquisite, spalted birch carefully honed by the lathe of his life. We don't need to know the full story of the spalt lines driven into his grain by the mold of war. We need only stand in respect for the fact that he has polished them so that his gift back to the world is beauty, wisdom, and humility.

Bud cutting a block of wood into size for the lathe. ⩗

Name: Kris Johnson

Occupation: Insect and disease specialist, U.S. National Park Service

Point of Interest: Exotic forest pests are imported from other countries and have no natural enemies in their new homes. They provide a tremendous challenge to land managers. And they remain one of the strongest arguments for wilderness with its natural species diversity and resistance to such invasions.

CHAPTER 5

The Tree "Doctor"

Kris Johnson

Forest diseases are one of the strongest arguments for

the preservation of forest wilderness areas.

—KRIS JOHNSON

IN EIGHTH GRADE, KRIS JOHNSON'S classmates predicted she would grow up to become a tree surgeon. "I used to spend my entire recess using Kleenex and tape to repair the damage sapsuckers were doing to maple trees on the playground," she says, chuckling.

Some thirty years later, Kris, though not a surgeon, is treating trees for disease and injury. As the supervisory forester for the Great Smoky Mountains National Park, Kris uses tools that are more sophisticated than tape and Kleenex, though she still brings the same quiet, steadfast passion to the care of the Smokies' millions of trees that she did to the dozen maple trees on her Tennessee playground.

Within the boundaries of the 540,000-acre Great Smoky Mountains National Park, great stands of indigenous southeastern American forest are protected from development. As America's most visited national park (nearly ten million visitors per year), it is renowned for its diversity of plant and animal life and the quality of its remnants of southern Appalachian mountain culture. Located partly in Tennessee and partly in North Carolina, the park contains some of the last remaining examples of old-growth forests in the East.

On a late March day, walking through a section of the park she loves, Kris is the consummate naturalist. She tucks her shoulder-length hair, just starting to gray, behind her ears and pauses to listen to the clear whistle of a tufted titmouse high in a leafless butternut tree above her. "He's busy setting up his territory, a sure sign of spring," she says with a smile. A short distance ahead at a swampy point in the trail, she points out the furled leaves of a trillium pushing their green strength up through layers of composting twigs, branches, and dead leaves. Kris pauses to inhale the sweet smell of earth returning to life after the long winter. If asked what other natural events she is thinking about on a spring day, Kris will share urgency about saving the forest from a nearly invisible foe.

"While early-season hikers like us are busy enjoying the mating calls of tufted titmice and cardinals and are looking to see treetop buds swelling into leaves, tens of thousands of tiny, aphidlike insects called hemlock woolly adelgids [pronounced a-DEL-jid] are busy hatching and setting out to destroy our eastern hemlocks." Kris goes on to explain that these hungry little creatures (about the size of the period at the end of a sentence) hatch from Q-tip-like egg sacks located on the upper branches of eastern hemlock trees high above the view of most people. Within two days, the swarms of adelgids have found the newest needles on trees and begin sinking their tiny mouthparts into these needles, inserting toxic saliva that both kills existing needles and weakens the trees' ability to produce new ones. Out of reach, out of sight, their presence is nearly undetectable until trees start to die.

Kris understands the intricate complexities of her forest and has devoted her entire career to preserving its beauty and confronting the challenges it faces from insects and diseases. The problems facing Kris and a permanent staff of six employees would discourage, maybe even immobilize, someone of lesser determination. She knows the odds. In 1962, the cousin of the hemlock woolly adelgid (HWA), the balsam woolly adelgid (BWA), was first discovered on Fraser firs in the Smokies. Since then, the BWA has killed 91 percent of all mature firs in the park. The newcomer, the HWA, was first discovered in the park in 2002. If this new creature wreaks the same

⌃ *Woolly adelgids identifiable on the needles of the eastern hemlock by their Q-tip like egg sacks.*

havoc on eastern hemlocks, another source of year-round greenery in the park will be gone forever.

"With the urgency of the situation, a talk on tree diseases can leave an audience totally discouraged," explains Kris. A gentle and scholarly introvert, she shows her passion in her love of earth and forest. "I want people to understand the severity of the situation, but I don't want them to feel hopeless. Tree diseases challenge us to understand how complex, resilient, and fragile nature can be. A catastrophic event, like the loss of Fraser firs, illustrates the critical importance of preserving wilderness where there is enough species diversity to combat large insect and disease outbreaks." Species diversity means variation within a given species. For example, some eastern hemlocks may have greater resistance to the attacks of hemlock woolly adelgids than others, in much the same way that some human beings have greater resistance to certain diseases than other human beings.

Early Background

Long before she knew a term like "species diversity" and what it meant, Kris seemed destined for a career that involved caring for wild creatures and places. She was one of six children in her family growing up in rural Tennessee. "I stayed outside until dark every day I could. My family hiked and camped, and I was active in Girl Scouts. I knew the names of trees because my father was a forester. And my interest in plants of the non-tree variety undoubtedly came from my mother's farming side of the family."

In college Kris expanded her curiosity to include the human world, so she majored in philosophy with a concentration in Asian studies. She focused on the liberal arts because she "simply enjoyed learning everything." With a degree in liberal arts, her job options were limited coming out of college. Her first job was working in an immunology lab as a technician. Her main objection to that job was being indoors. "People told me to 'grow up,' that the world of work wasn't about doing what I wanted to do."

But Kris didn't believe that philosophy and began to carefully chart the course of her life. Remembering that her first botany teacher, Ed Clebsch, had encouraged her to think about getting a fellowship to attend graduate school, she applied to study dendrology at the University of Tennessee. Dendrology is the branch of forestry that deals with the study of trees and shrubs.

From the moment Kris took her first dendrology class from Ed Buckner in 1976, she knew she had found her calling. "He was a marvelous teacher. He could read the history of the woods. I remember coming into a forested area that contained some tulip poplar trees. He knew right away that the area had once been a farm since tulip poplar is an early successional species. I wanted to understand the woods like he did."

As a graduate student, Kris decided to focus her research project on the BWA, the exotic insect pest that was killing Fraser firs. "I was an 'adelgidologist,' a rare species myself," smiles Kris. "I admit I mainly chose the project because it enabled me to backpack in the Smokies, but it also got me in on the ground floor of what would become a major focus of pest and insect research in the eastern United States. We visited Fraser firs all over the park to see if they showed evidence of adelgid infestation and to determine site and individual tree factors influencing infestation levels."

The Fraser fir is the only native fir in the southeastern United States, and it is a major component of all high-elevation forests in the park. According to Margaret Lynn Brown in *The Wild East* (University Press of Florida, 2000), 2 percent of the Smokies are covered with spruce-fir forest. These forests sitting on the tops of high peaks are alpine remnants from the last Ice Age.

At the time of Kris's 1978 study, many firs were infested but not yet showing decline. "As someone just starting in my field, I simply loved hiking into remote places. But, you know, I never dreamed we'd be the last people to see some of those stands." By 1982, the climb to the park's highest point, Clingmans Dome, was lined with naked firs. No one could imagine that hemlocks would fall prey next.

Forging a Career as a Tree "Doctor"

To get to her current position of supervisory forester in charge of insect and disease control in the Great Smoky Mountains National Park, the young adelgidologist had to work her way up the ladder. In the 1980s, many starting jobs in the National Park Service required a focus on law enforcement. Increasingly, national parks were becoming places for drug trafficking and crime evasion. It was the moonshine tradition many times worse. So, the athletic, quiet lover of trees spent a number of years carrying a gun, working in law enforcement at both the Chickamauga Battlefield and the Blue Ridge Parkway.

"I became a wiser, more tolerant, and confident person by meeting the cross section of society seen in law enforcement, and I helped a lot of people. I learned about criminals, people in trouble, emergency medicine, dark icy roads, and all types of automotive problems. I spent more time than I wanted in a patrol car, and I had lifesaving friendships with my fellow rangers. I saw thousands of visitors having

⤳ *The Great Smoky Mountains National Park received its name from the haze that mutes the rolling, forested mountains.*

Photo Courtesy: Ann Linnea

wonderful experiences in 'my' parks. I saw horrible accidents and gave people news they didn't want to hear. And there were also foxes in the moonlight, bobcats in the snow, and looking down on rainbows in the valley!"

Through dedication, hard work, and fortuitous job openings, Kris finally found her way into full-time natural resource work within the national park system. Now, as supervisory forester for the Great Smoky Mountains National Park, a large part of her work focuses on insect and disease control. "To speak about exotic—that is, originating from someplace else—forest insect and disease, I need to speak first of the chestnut blight. In the 1930s, the Asian chestnut blight fungus was introduced to this country on horticultural stock. It is difficult for us to comprehend the huge ecological, economic loss that one primitive organism caused."

One in every four hardwoods east of the Mississippi River was a chestnut. They were fast-growing trees that reached heights of one hundred feet and could have a trunk circumference of over thirty feet. One-fourth of the hardwood lumber harvested in Appalachian forests then was American chestnut, and their nuts were important to both humans and wildlife. "My parents' generation remembers that places like Yellow Mountain, Yellow Ridge, and Yellow Creek were all named after the chestnut's summer flowers."

This one species of fungus eventually killed nearly every chestnut within its natural range from southern New England to Mississippi. The loss was staggering. Kris's father, a career forester for Bowater Paper Corporation, watched the tragedy firsthand. When another Asian invader headed for Fraser firs, Kris was a graduate student adelgidologist at a crucial time.

Native Pests versus Exotic Pests—A Basic Understanding

With decades of practice explaining complex forest insect and disease issues to ordinary citizens, Kris begins most talks by outlining the difference between native pests and exotic pests. "Native forest pests like pine bark beetles are cyclic. They have always been with us on this continent. Therefore, these insects have native predators, and native trees have developed some resistance to them. The balance can hold unless we do something unnatural, like suppress naturally occurring fires or grow trees in monospecies plantations."

Wild, native pine stands are usually found on dry, south-facing mountain slopes. Prior to Smokey the Bear policies (implemented August 9, 1944), these stands burned regularly due to lightning strikes. Natural burns killed off pine bark beetles and the weakened trees that housed them, keeping native pine stands low in beetle infestations. However, with the advent of Smokey the Bear, most fires in national forests and parks were extinguished as soon as possible. This saved property and preserved the scenic view of trees but has turned out to be an ecological problem for the forests themselves. The advantage shifted from the native trees to the native beetles. Beetle populations began to rise, causing the death of significant numbers of pines. However, in the past decade, both the National Park

Photo Courtesy: Great Smoky Mountains National Park

Close-up of the exotic pest, Hemlock Woolly Adelgid. ⌃

Service and the U.S. Forest Service have begun to manage native pine stands with fire by letting lightning strikes burn unless structures or lives are threatened.

With the return of fire, one advantage has been taken away from the native beetles and returned to the native pines. Unfortunately, the creation of pine plantations has thrown the natural system of checks and balances back into crisis. A monospecies plantation is a giant tree farm where thousands of acres of one tree species, usually a pine of some type, are grown and harvested for paper products, plywood, and low-grade construction timbers. In the twentieth century, pine plantations in the mid-South were established by killing the native hardwoods with herbicides then planting monocultures of pine, often single genotypes of a fast-growing species not native to the region. Since every tree species has its own complex microbiology developed over generations on any given site, the insertion of farmed pines into recently poisoned soils with no microbial link further added to the fragility of the crop. Thus, these tree farms were basically established on sites poorly adapted to pine, creating entire plantations of stressed trees—perfect conditions for pine bark beetle infestation to build to epidemic proportions. The pine bark beetle (which thrives on any type of pine) now threatens these tree farms almost as though it were an exotic pest—no resistance in the trees, no natural predator, no fires.

Exotic pests, unlike native pests, do not originate in the ecosystem where they are causing the problem. They are most often inadvertently imported on garden and houseplants brought to this country to appeal to the suburban landscaper. Once here, these insects or diseases generally have no natural predators and easily begin reproducing and moving off their host plant to find other hosts. This is what happened when the chestnut blight fungus arrived in the United States in the 1930s. It undoubtedly arrived on an Asian chestnut tree, which had resistance to it. However, once it began reproducing and found its way to the American chestnut, almost nothing could stop its progression. Foresters cut down infected trees to try to create a transmission barrier, but eventually nearly every chestnut tree in the eastern United States died.

The chestnut blight should have been a signal of more problems to come. "One logical solution would have been to legislate and enforce interstate and international quarantines on nursery stock, but that's not popular. That is, it's not good for short-term business. Unfortunately, too often little attention is paid to long-term economic impacts," Kris explains. So, the gardening industry keeps importing exotic plants from Asia, Europe, Africa, and South America, and each imported plant represents a potential free ride for insects and fungus that have no natural enemies in North America. In addition, many insects (particularly wood borers like the Asian long-horned beetle and emerald ash borer) and diseases can travel in packing material, pallets, firewood, furniture, and other wood products.

The next exotic pest to affect the Great Smoky Mountains National Park after the chestnut blight was the balsam woolly adelgid. Arriving in 1962, the tiny insect—one-twenty-fifth of an inch—began destroying the Fraser fir. The insect arrived on nursery stock imported from Europe and Asia, and like most exotics had no predators on this continent. As Brown notes in her book, "It sucked nutrients from living wood cells and injured them with a substance in its saliva. The substance caused abnormal cell division in the fir's cambium that resulted in hard, brittle wood, no longer capable of transporting nutrients." It took from three to nine years for an infested tree to die.

As it had done with the chestnut blight, the Great Smoky Mountains National Park attempted to minimize dispersal by locating and cutting down affected Fraser firs. Several areas were cut near Mt. Sterling, a method sometimes successful in controlling pine bark beetles but, as is now known, never effective with adelgids. In 1975, Canadian researchers found that the fatty acids in a simple soap solution killed the insect by dissolving its outer shell. The soap had to be dispensed with a high-pressure hose. That method worked to save Fraser firs grown as Christmas tree row crops, but it proved impractical in the remote conditions of the park where trees grow away from road or trail access. Seventy-four percent of all spruce-fir forest in the southern Appalachians is located in the Smokies. By 1978 (two years after Kris's graduate research), all of it showed signs of the adelgid infestation.

The death of the Fraser fir at high elevation began to change everything about the environment of the spruce-fir forests. Dead trees created an opening for more sunlight, which in turn killed mosses and lichens and allowed dry-loving plants, like blackberries and elderberries, to thrive. The change of understory plants affected nesting birds and native insects. Scientists are still studying to understand the ecological impact of that one exotic pest—and now a new pest has arrived.

Hemlock Woolly Adelgid (HWA)

Kris wasn't even born when the great chestnuts of the east went down. She was a young woman barely launched on her career when Fraser firs began to succumb. Now she is a midlife woman with skills and authority, and yet another species-threatening infestation has arrived in her park. Kris is both calm and concerned about the 2002 arrival of the hemlock woolly adelgid. (By 1992, most of the hemlocks in Shenandoah National Park in nearby northern Virginia were infested, so scientists knew it was just a matter of time before they reached the Smokies.) She is calm because that is her demeanor when facing challenge, and she is concerned because she has a lifetime of experience that informs her how deadly tiny critters in vast numbers can be.

"In Eurasia where the balsam and hemlock woolly adelgid originated, they are kept in check by a combination of environmental factors: predators, parasites, and host resistance. Here the BWA and HWA, no bigger than the dot on an 'i,' reproduce parthenogenetically [without sex] about four times a year, have no effective predators, and are readily spread by wind currents, birds, and people. A single female can generate as many as ninety thousand copies of herself in a single year. Once they arrived on imported horticultural stock, it was only a matter of time before they found their way to native trees. But we were still surprised by how quickly it got to the park."

≽ *Distribution map of eastern hemlocks within the Great Smoky Mountains National Park.*

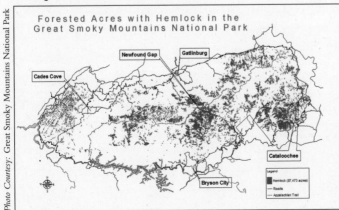

Photo Courtesy: Great Smoky Mountains National Park

In her preferred role as naturalist, Kris invites listeners to see the larger picture—to appreciate the beauty and role of the hemlocks before focusing on the critters that are attacking them. The park's hemlocks are primarily located in dense stands in remote valleys called coves. These areas (about one-fifth of the park) were never logged because hemlock, usually full of knots, is considered a poor lumber tree. Often referred to as the redwood of the East, the eastern hemlock provides deep shade with a rich understory of plants that keeps streams clear and clean. Mature eastern hemlocks can top 160 feet in height, be more than 6 feet in diameter, and live longer than six hundred years.

The HWA affects its host species, eastern hemlock, differently than the BWA affects its host species, Fraser fir. The HWA feeds on sap at the base of the hemlock needles. To bring this concept to life for third graders, Kris developed the idea of giving the children soda straws to mimic the action of the adelgids sucking juices from a very specific point. The adelgids (and the third graders) halt the flow of nutrients to and from the foliage. This basically starves a tree to death, causing needles to turn from dark green to gray and then fall off. And unlike the BWA, which feeds primarily on mature trees, the HWA affects all age classes.

However, Kris does not dwell on the negative. In the same breath that she is carefully recounting the awesome and awful force that these tiny creatures represent to the beautiful forests she is dedicated to protecting, she speaks hopefully about species genetic diversity. Genetic diversity means that two eastern hemlocks may look alike—in the same way two blond, thirty-year-old women look alike—but each of the trees (and each of the women) has different strengths and weaknesses. In much the same way that one young woman may be susceptible to diabetes because her parents had the disease and the other woman is not susceptible, one hemlock may have more resistance to the HWA than the other.

Years of working with trees and their diseases have taught Kris both humility and hopefulness. "Thirty years ago people said we'd never see another Fraser fir alive. But some stands have survived the onslaught of balsam woolly adelgids for reasons we don't understand. And even though the other deadly Asian exotic—chestnut blight—killed all mature trees in the eastern half of the U.S., the species is not extinct. Old stumps in the park continue to send up sprouts that grow into small trees before the blight kills them. Our researchers are trying to find sprouting trees big

> ### Kris's HWA Philosophy #1
>
> "People expect science to have answers to problems, but there simply are not answers to some of these problems with exotic species. With HWA we are trying some of the same techniques that we did with the BWA. However, once something like the arrival of the HWA happens, it's like a Pandora's box. As I said, the critters are tiny, they have a prolific life cycle, and they are easily dispersed."

enough to produce viable seeds for possible crossbreeding with blight-resistant Chinese chestnut trees."

In addition to protecting the future of the park as a repository of genetically diverse trees, Kris and her staff are working to find biological controls for the invading HWA. In 1998, Kris traveled to southwest China with a U.S. Forest Service research team to collect HWA predatory beetles. Her original degree in philosophy and Asian studies paid surprising dividends.

Kris's HWA Philosophy #2

"If there is enough genetic diversity in the park hemlocks, they might be able to survive. The park represents eight hundred square miles managed mostly as a wilderness. There is a large enough gene pool that there may be some hemlocks resistant to the HWA."

"It was a nice background to have when I went to China. It certainly enriched the trip for me personally, and my Chinese colleagues enjoyed my appreciation of their history and culture. I'm spending my forestry career fighting a Chinese beetle, and my background in Asian studies helped remind me that the HWA threat is not the fault of the Chinese people. The HWA is the wrong bug in the wrong place, so I went to discover and import its natural foe."

The tiny predator Asian beetle that Kris collected in China has been dubbed the St beetle (*Sasajiscymnus tsugae*). The St beetle is a black ladybird beetle about the size of a poppy seed that feeds on all life stages of the HWA. It preys only on adelgids, but unlike the adelgids, it must mate to reproduce, so its population does not increase as fast as the population of the adelgids. The St beetles (and most other biocontrol agents) are most effective when their prey populations are relatively small but robust. Presently there are three beetle-rearing facilities in the United States, with more being planned. Funds for much of the research comes from the nonprofit organization Friends of the Smokies and the U.S. Forest Service. By 2009, Friends of the Smokies had raised over $1 million to help with HWA eradication. Other predatory beetles, including some obtained from the Pacific Northwest, are showing promise as woolly adelgid eaters.

Go Get 'em Little Buggers, and Other Techniques

In April 2004, Will Blozan and two of his Appalachian Arborist employees climbed hemlocks in the Joyce Kilmer Memorial Forest to release some of the first St beetles to combat the HWA. The Joyce Kilmer Memorial Forest (located thirty miles southeast of the park) is one of the very few remaining large tracts of virgin old-growth forest in the Appalachians outside the park, and it contains many trees, hemlocks included, that are twenty feet in circumference and over four hundred years old.

Blozan and his colleagues released two thousand beetles near the top of three trees (125 to 140 feet tall). According to a press release by the U.S. Forest Service, a number of the old-growth hemlocks in Joyce Kilmer are already heavily infested and will likely be killed in a few years. Blozan's crew placed a beetle-covered branch into the top of each tree. The hope was that an upper canopy release would help in the distribution of the St beetle over a wide area. Blozan noted that many of the beetles flew at once. "I could not help feeling hopeful for the hemlocks, and was excited by the 'spunkiness' of the beetles and their immense task ahead," wrote Blozan. "Go get 'em little buggers!"

According to Kris, there is hope that the St beetles will save some of the park's eastern hemlocks. The big challenge is time and money. "We get backed into estimating the exact cost of this operation, and we really don't yet know how the St beetle will spread, reproduce, and feed. Generally, biological controls work best when the prey populations are lower. Raising St beetles in captivity is very labor intensive because there is no artificial diet—they eat only adelgids. Publicly funded facilities at the University of Tennessee, Clemson University, and North Carolina State University are currently raising the St beetles."

Any effective response to exotic pests needs to occur on several fronts. The park is also using ground techniques to combat the HWA—spraying campground and picnic area hemlocks with insecticidal soap and injecting pesticides into selected individual trees. As was true for the BWA, the HWA have no shell, so soap acts as a desiccant and kills them immediately. Foliar treatments are sprayed from truck-mounted spray units that can spray up to eighty feet next to roads. The limitation here, of course, is that trees must be near roads and less than eighty feet tall, and the application needs to be repeated every six to twelve months because it kills only adelgids present on the tree at the time of application.

Another technique being used to combat the HWA is the application of systemic insecticides. These insecticides are applied only to the tree root zone to minimize damage to surrounding organisms and move slowly up the trunk and out into the foliage. As an artificial kind of nicotine, the chemical kills the adelgids and apparently does not climb up the food chain or enter watersheds. Results have been dramatic and can last up to five years.

In a test plot in the Cataloochee Valley in the southeastern tip of the park, a soil application of 75-percent-concentration Imidacloprid was applied to ten trees in

November 2002. The valley contains more than 80 percent of the world's tallest eastern hemlocks. In the spring of 2003, Will Blozan was hired to climb each of the ten trees to collect foliage samples. Analysis of the samples revealed that some Imidacloprid was present and that the trees were successfully surviving. In June 2005, the ten trees were again climbed and sampled. Blozan and his climbers found the trees essentially clean of living adelgids and exhibiting various degrees of recovery with seemingly random limbs green and full growing right next to gray, decaying limbs.

"The apparent delay in chemical activity and the light to moderate infestation at the time of treatment may explain the patchy recovery," wrote Blozan in an internal memo to the Eastern Native Tree Society.

Photo Courtesy: Great Smoky Mountains National Park

⌃ *Using a foliar soap spray, a forest technician treats a young roadside hemlock for HWA.*

"Typically, untreated trees surrounding the ten treated trees were gray and declining rapidly . . . I know Imidacloprid works. I use it almost daily against HWA and have had excellent, albeit delayed results."

Neither Blozan nor Johnson advocates chemical treatment as the only approach to saving the eastern hemlock, partly because application is labor intensive and partly because affected trees must remain on the drugs indefinitely to survive. "Will describes it best," explains Kris. "We have a forest on life support."

The greatest successes of the HWA reduction program have been at Big Creek, Greenbrier, Albright Grove, and most of the developed zones along roadsides, picnic area, and campgrounds. Other areas of the park that have been treated that seem to be holding their own are the Roaring Fork Motor Nature Trail and the entrance from Gatlinburg. "We had hoped to save more trees in the Cataloochee area where most of the hemlocks are concentrated," Kris says, "but the massive concentration of hemlocks also contributed to the massive buildup of adelgids, and we were dealing with drought,

⌃ *Application of systemic insecticide is done at the root level by a trained technician.*

which limits a tree's ability to take up the chemical." Another area of notable devastation is visible from U.S. 441, the Newfound Gap Road. "What people see is pretty shocking and depressing. The slopes were too steep, the soils were too thin, and the trees too remote for us to treat them." So, now the Gap is covered with gray and ghostly spires.

Insecticides buy time while other cures, such as predatory beetles, are being tested. The sheer size of the Great Smoky Mountains National Park raises hopes that there may possibly be genetically resistant individual hemlocks. Finding genetically resistant individuals or pockets of trees cannot save the eastern hemlock as a predominant climax species, but it may preserve remnant populations for eventual rebound.

Spiritual Warrior in the Bureaucracy

In a rare moment of rest, Kris leans against the trunk of a hemlock the way others might lean against the shoulder of a friend. "The Smokies is a great park. I love my job. Every day something new and exciting pops up. I get to mentor my seasonal employees—and I have the opportunity to start people on careers that may mean as much to them as mine has to me. Even though I may not always have the resources I need to do my job well, I have tremendous autonomy—and constant challenge. I am one of the few people my age who goes off to work with a day pack and loves her job.

"If I approached my job as a bureaucrat, if I focused on budgets and people management, I'd be caught up in meetings, in forming task forces, and in reporting results to more meetings. Nature teaches me a different way. Nature teaches me to focus on the importance of trees to the world—for oxygen, for water quality, for aesthetics, for habitats. We can't make it without trees."

Whether conducting a forest regeneration survey, managing budgets, or mentoring staff, Kris Johnson loves her job. »

Kris balances the considerable rigors of her job with time for music, gardening, and, of course, walks in her beloved eastern woods. "My business and my hobby these last eight years has been to be in the hemlock forests as often as possible. It's been important to me to see and honor them." She is an avid reader both for pleasure and for her profession, and she plays flute in the local community concert band—one of her greatest stress relievers. And she is keenly aware of interconnected environmental issues. She allows herself to be discouraged but not immobilized. Her steady manner belies the tremendous determination and genuine activism that has charted the course of her life.

"We can't control nature, nor should we. Human activity introduced these diseases and pests, and I'm trying to undo some of that damage with informed human intervention. But in the end, it's up to the trees: Can they develop resistance in time? As we study nature's response to forest diseases, we are reminded of the importance of the preservation of forest wilderness areas. The trees give me an opportunity to be a spiritual warrior for the things I love."

Photo Courtesy: Great Smoky Mountains National Park

Photo Courtesy: Ann Linnea

Name: Will Blozan

Occupation: Arborist

Point of Wisdom: Large, old trees inspire reverence. Finding and measuring them provides crucial forest restoration information.

CHAPTER 6

The Arborist and Big Tree Hunter

Will Blozan

We will turn down work if it's not the best thing for the tree.

—WILL BLOZAN

SWINGING FROM A CAREFULLY rigged climbing harness, suspended from an eastern hemlock, Will Blozan speaks about the trees he loves so dearly. "Eastern hemlocks deserve more than being wiped into historic oblivion. I am passionate about making sure we never forget them."

Will is passionate about a lot of things. In his early 40s, he is already president and cofounder of the Eastern Native Tree Society (ENTS) and owner and founder of Appalachian Arborists, Inc., in Asheville, North Carolina. As one of the prime forces behind saving eastern hemlocks from impending extinction, he is widely acknowledged as one of the prominent eastern big tree hunters. At a trim and fit six feet, two inches tall, he is a father and husband, an adventurer, an organizer, and a scientist. He looks happiest wearing a helmet and climbing harness and hanging from the limb of some huge tree.

Unlike big game hunters, big tree hunters are not out to kill their quarry. They are out to save them. Big tree hunters are a rare breed. Some hang out in trees to satisfy their own curiosity; others undoubtedly revel in the opportunity to get their names in the record books. Will's interest lies much deeper.

"Big trees inspire awe, even reverence in me. And by getting accurate information on where and how they grow, important scientific snapshots can be gained to help with future restoration."

When Will speaks about restoration, his current focus is primarily on documenting the eastern hemlock. This lacey conifer is best known in the southern Appalachians as the elegant tree that shades trout streams and flowering rhododendrons. Eastern hemlocks can measure over 19 feet in circumference and over 160 feet in height. In one of the better-known groves of old eastern hemlocks, Joyce Kilmer Memorial Forest, some trees are over four hundred years old. Unfortunately, as noted in the previous chapter, nearly all eastern hemlocks are currently under siege and quite possibly en route to extinction in the wild.

On a December 2003 climbing expedition, Will recorded his concern about the eastern hemlock in a memo to members of ENTS. (Will and four colleagues founded this organization in 1996. Playing with the name of the fabled tree stewards in J. R. Tolkien's famous trilogy *The Lord of the Rings,* they came up with an acronym that describes their scientific fierceness for accurate description and location of large eastern trees.) Since that time, Will has been intimately involved in state, regional, and national efforts to save eastern hemlocks. Will recently flew to California to present to the executives of a Japanese executive chemical company whose product is used to save hemlocks on the brink of extinction. During the question and answer session, he uttered the now famous line directed toward those not working to save some hemlock groves for the future: "I have no sympathy for apathy."

Will's December 2003 ENTS Memo

Last Gasp of the East Fork Hemlocks

Yesterday Jess Riddle, Ed Coyle, Mike Riley, and I went into the unsurpassed tall hemlock forests of the East Fork of the Chattooga River in the Ellicott Rock Wilderness, South Carolina. We braved an impending ice storm, a swollen river crossing, and ice-covered logs. A traverse was set up with rope and pulley to get the gear across the river. Then we went to climb a tree I had measured two years ago after climbing the "East Fork Spire," a 167-foot 10-inch eastern hemlock that was a new world record height (living tree) for the species. Two years ago I had measured the new tree (still unnamed) at 168 feet 11 inches with my laser-clinometer measurements.

Once to the site, Ed used his new laser-clinometer equipment to remeasure the tree and got a height of 168.92 feet. I then used my equipment and got a measurement of 168.95 feet. Not bad!

So, up we went! The tree was 114 feet in girth, and the tape drop indicated a height of 168.9 feet. A new world record! Upon seeing that there was only two inches difference between our ground laser-clinometer readings and our tape drop, Ed asked, "Why do we bother to climb?"

Well, part of the reason is in the title of this e-mail. This tree and every other in the grove are essentially dead. This was the last climb of the last tall hemlock in the grove. The hemlock

≈ *Will Blozan in the Usis hemlock, Great Smoky Mountains National Park, Cataloochee, NC.*

Photo Courtesy: Jason Childs

woolly adelgid has destroyed the grove, and some trees are completely defoliated and dead. The tree, which I had climbed just two years ago, showed no obvious sign of hemlock woolly adelgid damage at that time. Now it is a partially defoliated, gray ghost of its former luscious glory. We are documenting these great trees down to the very inch of growth literally on the eve of their death. We are also looking for answers to nagging questions.

Can eastern hemlocks reach 170 feet? I would have to say, absolutely, but not any more. The last growth increment on the tree I just climbed was 7 inches long. The top and every other tip on the tree are now dead. This tree and the East Fork Spire would have been 170 feet within three to five years, if hemlock woolly adelgid were not a factor. I don't know how tall they would have grown to, but the tops of these trees were arrow-straight and vigorous, with no sign of slowing down. Unfortunately, we will never know. Eastern hemlock is a clear example of a species lost in its prime.

Time is out folks. The great hemlocks of the southern Appalachians will likely be just a historical anecdote in a few more seasons. At least, because of ENTS, the history will be accurate and truthful.

The story of how a tiny insect like the hemlock woolly adelgid can wreak such havoc is explored in greater depth in Kris Johnson's story, in the preceding chapter. This chapter introduces us to the precarious, passionate work Will and his cohorts are doing on behalf of large eastern trees in the United States. They have been part of the clarion call to document hemlocks, and they have been part of the fight to try to save them. Blozan and other ENTS have spent about a hundred thousand dollars of their own money identifying and measuring the world's tallest and largest eastern hemlocks through the Tsuga Search Project. (The scientific name of the eastern hemlock is *Tsuga canadensis.*) And Will has personally been hired by the U.S. Forest Service, the South Carolina Department of Natural Resources, and the National Park Service to do much of the chemical and biological treatment of eastern and Carolina hemlocks—treatment that must be handled by skilled arborists willing to traipse deep into wilderness areas.

An Arborist

Big tree hunting is Will's avocation. His vocation focuses on trees of all sizes. As a professional arborist, he works to improve the quality of tree care in western North Carolina. In 1998, at the age of thirty-one, he started his own business, Appalachian Arborists, to provide tree care in Asheville, North Carolina, and support his young family. He already has five full-time employees with additional summer seasonals. "Our job as arborists is to keep trees healthy and happy for the people who own them. Unlike forestry, it's a very people-oriented business."

Photo Courtesy: Ann Linneas

⩘ *Will organizing his gear for a day's work as an arborist.*

"Our reputation is very good. We will turn down work if it's not the best thing for the tree." Will tells the story of a customer who called to get an estimate for cutting down seven trees in an upscale housing development in the rolling hills surrounding Asheville.

"When I got there, I couldn't believe my eyes. The trees were breathtaking, absolutely spectacular! I asked the man why he wanted to cut his trees down. He said the two tree companies who had previously appraised the site claimed that some of the trees were too crooked, others were too big.

"Well, I took a deep breath and as calmly as I could explained that the tree that was too crooked was a native sourwood nearly 250 years old—and that sourwood grows in a crooked fashion. [Sourwood, a tree-size relative of blueberries, is an understory tree with an unusual twisted, leaning growth habit.] I also explained that a couple of his hickories were nearly state records. The *only* tree I would remove was a smaller oak leaning toward his house. He was happy—I saved him money and gave him a whole new appreciation for his trees. I was happy—I saved six spectacular trees."

Will's path from a boy who climbed trees for fun to a man who climbs trees for a living (and also still for fun) reads like a chronology of what to do to become a professional tree climber. Yet, by his own admission, he didn't really plan it that way; his work just evolved. "As a kid, I remember always being impressed by trees, always climbing them. My brother and I had separate trees in the backyard. As we got older,

The Arborist and Big Tree Hunter   91

Dad put a piece of tape on the trunk of each of our trees letting us know we shouldn't climb any higher than that mark. I think that's when my desire to push the limits really got started. I'm not sure, but I think I even moved the tape up.

"My parents were incredibly supportive. They sent me to youth wildlife camps, and after high school I spent nine months traveling with the Audubon Expedition Institute. [Offered in conjunction with Lesley University, this program has been offered for over thirty years and is designed to offer field learning for students and faculty.] Twenty-four people traveling in a school bus turned out to be too much closeness for me, but I learned so much, and it gave me a goal: One day I wanted to be one of the resource people that bus came to visit."

After the Audubon Expedition Institute, Will returned home to Maryland and got a job with a family friend working for a tree service. "Wayne Anderson really formed the rest of my years," says Will. "He was passionate about keeping trees happy, healthy, and alive for people. And he was your total archetypal good worker. He had an incredible sense of productivity and efficiency. First I apprenticed on the ground. Then he taught me how to use ropes safely. I learned good techniques right from the beginning."

College Years

In January 1987, the young man from Maryland was admitted to Warren Wilson College in Asheville. "I chose the school because it had big, beautiful trees and because participating in work crews was part of the curriculum. It was the best education I could have had."

Will's first work program experience was as a janitor, but he soon managed to get assigned to landscaping, where he exercised his strong leadership skill. "I had noticed that the trees on campus needed a lot of hazard reduction work. Thanks to my experience with Wayne, I had the confidence and skill to make a proposal to the college for a tree maintenance program. They assigned me three students, a truck, a yearly budget, and an equipment allotment.

"Now when I return to campus, it makes me feel great to see that program still going. In fact, I regularly recruit Warren Wilson graduates for my business.

"But Warren Wilson was about more than just the tree crew. I took every class that had anything to do with the environment—ecology, taxonomy, and wildlife management. One of the best classes was plant taxonomy taught by Gary Kauffman [now a U.S. Forest Service botanist]. He was passionate and clever, a kind of scratch-and-sniff teacher. We'd be walking along, and he'd explain that a certain plant should be in this vicinity because of habitat. All of a sudden he'd dive under a bush and come up with that plant. He was totally amazing."

One of Will's college projects was "Roadside Dendrology from Maryland to Nova Scotia." He marked the mileage points where tree species changed and then did an overlay of how soils and geology changed correspondingly. Even though his sampling was done along totally disturbed roadsides, he could see the intricate transition from deciduous hardwoods to spruce-fir forests as he drove north. It gave him great comfort to see that the grand scheme of nature held despite the encroachment of civilization.

After college, his love of exploration and trees took him to the former Dutch Guiana (Surinam) in South America. There he did trail work for six months, helped with bilingual signage for an arboretum, and pruned trees on fruit plantations. Living and working in the great forests of the next continent south, Will further honed his ability to see everything as part of a continuum. The following spring he started work in the Great Smoky Mountains National Park and was impressed by some of the habitat similarities to Surinam. "The sweet smell of decaying logs was no different from the smell of decaying logs in the tropics. And the fact that there were scarlet tanagers in both places amazed me in a way that reading about it in a book never could have."

In March 1993, Will was hired to work on an old-growth inventory project in the Great Smoky Mountains National Park and moved to Gatlinburg, Tennessee. "We tried to use existing protocols for determining whether a stand was old growth. Then we were to map the old-growth stands in the park. There were four rating scales in eight criteria—age, size, amount of debris on the ground, and so on. A score of sixty represented a perfect old-growth stand." However, Will found that in the real world of the forest, there were more exceptions than rules.

"For example, let's look at a small pocket of old growth in the park. An ice storm hits and pretty much wipes it out. Is it still old growth, even though there are no old trees anymore? Of course, it is! It's just at a different stage of development. One of the things we found, of course, is that big trees are not always old. And small trees are not always young. There are species that at age four hundred you can reach your hands around the trunk. Old-growth forests simply don't fit in boxes." Though Will lost patience with debates over what was or was not old growth, he gathered and categorized much valuable information while he was a project technician.

Those years in the field also introduced him to other scientists and technicians who would pave the way for his Appalachian Arborist company to be involved in attempts to stop the onslaught of hemlock woolly adelgids. And, of course, the project really launched him into his avocation as a big tree hunter. "What I got out of that project was the privilege of seeing lots of huge, old trees and realizing my dream of serving as a resource person for the Audubon Expedition Institute. They came and spent a day with me while I was a technician on the old-growth project. I just slipped easily back into that circle. It was great for me."

Searching for the Big Ones

In Will's first three years on the old-growth inventory job, he discovered sixteen national champion old-growth trees. Listening to him speak or write about a tree climb, one can tell that the sheer adventure of finding and climbing tall trees in remote places in bad weather is obviously part of the fun. His earlier ENTS's memo mentions an impending ice storm and crossing over icy logs on a swollen river. Descriptions of other climbs involve feats like carrying eighty pounds of climbing gear nine or ten miles from the nearest road or swaying precariously from a treetop perch nearly two hundred feet from the ground.

Ordinary people would believe that every inch of forest in the United States has been traversed numerous times over. How is it possible that people like Will keep "discovering" something as large as 160-foot-tall, 15-foot-girth trees? It takes someone with a trained eye to recognize a truly big tree, and few people would venture into some of the places Will seeks to find them.

On March 31, 2006, Will and his crew went in search of a hemlock potentially larger than the world record tree discovered in his December 2003 memo. A U.S. Forest Service specialist had reported the new tree in the Kelsey Tract near Highlands, North Carolina, to Will during an aerial predator beetle release. The forest in this area begins at 3,800-feet elevation and goes up to 4,260-feet elevation. Dense shrub layers of dog-hobble, rosebay rhododendron, and cinnamon clethra clog the understory.

The challenge to Will is not only finding large trees but also accurately measuring them. To understand this determination to be accurate, one must understand a bit about the politics of recording champion trees. Since 1940, American Forests, the nation's oldest nonprofit conservation organization, has published the *National Register of Big Trees*. To be eligible for the champion tree registry, a tree must be native or naturalized to the United States. Points are given for every inch of trunk circumference, foot of height, and a quarter of its crown spread in feet. Measurements have been done with a tape measure and a clinometer (determines angle of inclination). Researchers have extended a tape measure around the tree for its circumference and

All in a day's work when hunting for big trees. »

then measured off a distance one hundred feet away from the trunk to "shoot" an angle to the top of the tree with their clinometer. The tree's height is figured using that angle reading and a simple mathematical formula. But Will, other members of ENTS, and Dr. Robert Van Pelt, a forest ecology researcher for the University of Washington (see Chapter 10), are challenging the assumptions behind that method.

"Traditional techniques assume that tree trunks are straight and the high point of the crown is evenly situated over the base," explains Will. "That assumption is insanely inaccurate. When does that actually happen in nature? The main crown of a tree can be as much as fifty feet off the center of the trunk. Trees in nature seldom grow straight up.

"We use a measuring tape, a clinometer, *and* a laser rangefinder. We take the laser reading from several vantage points. It actually reflects off the top of the tree, measuring the tree itself—not extrapolating some fairy-tale version of what the tree looks like. And then, just to double-check, we often climb the tree if it appears to be a new all-time record."

In the story of the East Fork hemlocks, Will and his climbing partner, Ed Coyle, determined that there was only a two-inch difference between their laser-clinometer readings from the ground and their measurements from the top of the tree. Will's answer to the question about why they go to the trouble of climbing trees if ground measurements are so accurate has two parts. First, they can more accurately see how a tree is growing from up high. And second, they can verify their new technique.

"The record books are full of inaccuracies," says Will. "We are challenging the status quo, and not everyone is happy about that. We have gotten some unbelievable resistance from PhDs and foresters who claim it's been done in a certain way for decades, and they have the scientific literature to back them up.

"I have sixty champion trees I could nominate from North Carolina alone," continues Will. "But at the moment, the powers that be are saying those nominations have to be checked by traditional methods. All I know is that ENTS is amassing a huge, accurate database of what eastern trees are really capable of, *and* we're offering tree-measuring workshops that American Forests's staff are attending. Time will tell if our work will be accepted by traditional sources."

Eastern Hemlock Record by New Standards

The Cheoah Hemlock near Highlands, North Carolina, proved challenging both to approach and to measure in March 2006. In Will's words to ENTS members:

« *Will hanging out on the end of a branch preparing to do measurements for total tree volume.*

For some reason this tree had an unbelievably complicated crown dominated by multiple splits, forks, and reiterations. This complex architecture added to the climbing experience in many ways, not the least of which was simply being surrounded by an utterly artistic and strikingly beautiful tree that literally engulfed and humbled us. The trunk sections were so huge and the canopy so dense it was as if we were climbing in a huge head of broccoli.

Attaching a tape from the tip of the leader to the bottom of the trunk, Will and crew took six hours to map and measure the complex tree, which had four tops: 151.7 feet, 155.7 feet, 156.5 feet, and 158 feet. By the time they added up volumes for all the multiple tops and reiterations, they estimated the tree at 1,563 cubic feet—11 percent larger than any other hemlock documented:

Folks, barring the very slim chance of some unknown discovery of an immense hemlock elsewhere in the southern Appalachians, I feel we have found the largest known specimen of its species. . . . The largest Great Smoky Mountains National Park contender we all agreed was the Caldwell Colossus, which is fully 12.1 percent smaller than the Cheoah Hemlock.

The disappointing part of this discovery is that the hemlock woolly adelgid had already attacked and killed all the terminal tips on the Cheoah Hemlock, despite the release of predatory beetles two years earlier. However, Will's drive to act fast and save some of the hemlocks paid dividends in this case. He persuaded the ranger district to spend funds on insecticidal intervention, and when Will visited the tree in fall 2009, it was "doing well—currently the largest and tallest *living* hemlock." The first treatment with a biological control (predatory beetles) hadn't worked, so chemical intervention was essential to save this particular record tree.

The decision of whether to use predatory beetles or systemic pesticides is not a simple one to make. The systemic insecticide must be hand applied when the trees are in good vigor, the soil type is well-developed humus, and the moisture conditions are good. The chemicals largely remain soil bound and are absorbed through drenched soils into the tree's vascular system. Land managers knew the chemicals stopped hemlock woolly adelgids. They didn't know if soil microbes would be affected or if any of the chemicals would make their way into the watershed. They didn't have time to wait for those studies. Hemlock woolly adelgids are so prolific and so fast acting that only quick chemical treatment could save any of the wild eastern hemlocks. They took the risk.

In his inimitable style, Will wrote to the ENTS:

We have already lost the tallest known eastern hemlock forest ever documented—the East Fork Grove in the Chattooga River watershed in South Carolina. This grove contained six

As part of the extensive Tsuga Project, Jess Riddle is running tape for vegetation plots on the fourth tallest eastern hemlock, Noland Mountain, Great Smoky Mountains National Park, Cataloochee, N.C. »

of the eight known trees outside of the Smokies over 160 feet. They were destined to grow to exceed the record height of 169 feet 10 inches within five years, but the sucking bastards got to them first.

The Tsuga Project—Documenting and Preserving Superlative Eastern Hemlock

Will does not give up easily. When he began to see the tidal wave of tiny black insects overtaking his beloved hemlocks, he went into immediate action. "From the beginning I knew how serious this pest was and how quickly it kills. I set two goals: preserve as many of these magnificent eastern giants as possible and document their grandeur for future generations if they were to be lost."

Building on a background of working with the Great Smoky Mountains National Park and his big tree climbing techniques and expertise, he and fellow ENTS members created a proposal to gather benchmark data on a species and its environment on the verge of ecological extinction. With the scientific advice of Dr. Lee Frelich (University of Minnesota), Dr. Robert Van Pelt (University of Washington), and Dr. Stephen Sillett (Humboldt State University, California), Will put together a funding proposal to the Great Smoky Mountains National Park to locate, document, and preserve exceptional specimens of eastern hemlock within the park and adjacent lands. Ninety percent of all the largest and tallest eastern hemlocks in the east are located in the park.

Will approached his company, Appalachian Arborists, and asked if it would support this two-year, intensive project, which would find him mostly gone from day-to-day operations so he could climb and study the remaining hemlocks before they died. He got the company's support, and he got a grant from the park. It was an epic, intensive two-year project of sometimes brutal backcountry hiking and climbing. Working from September 2005 to October 2007, Will and Jess Riddle identified seventy-five "superlative" trees (those over 160 feet in height). Fifty-three of these trees were located as a result of this project; twenty-two had been previously located by ENTS. "The vast majority of the new finds were growing on Big Fork Ridge, NC-GRSM (Great Smoky Mountains National Park) on or presumably in the alluvium of Roaring Fork Sandstone. Of the seventy-five tallest hemlocks, only four were located on the TN side of GRSM. Of the remaining seventy-one trees, five were in SC, one in GA, one on private land in NC, and the rest—an astonishing 90.1 percent—were in Cataloochee Valley, NC-GRSM."

Forty-two of the superlative hemlocks, located in five states, were climbed. From the final abstract, "The big trees were climbed, measured in detail, and modeled for total displacement volume. The fifteen tallest and largest were selected for further study. All trees grew on northerly facing slopes in near-temperate rain forest conditions between 2,160 and 3,900 feet in elevation. The tallest tree measured was 173.1 feet tall."

The 120-page report contains photos, biomass detail, surrounding forest composition, vegetative plot results, height and volume measurement, slope aspect, and elevation. "It is a stunning report," says Kris Johnson, Great Smoky Mountains National Park forester. "Thanks largely to Will's dedication, we have a record of the eastern hemlock that will be forever valuable."

Will and his technician, Jess, did all the reconnaissance and then went back to climb, do all the plot work, and treat the trees that had a chance of being saved. "It's clear we have lost the best of the giants," says Will. "I wasn't alive for the demise of the American chestnut. If I wanted to find information about those magnificent trees, I can't. They weren't measured. I did not want eastern hemlocks to go into this kind of historic oblivion. They deserved more than to be wiped off the face of the earth."

Will spent more than $100,000 of his own money and that of Appalachian Arborists. Unfortunately, not everyone is as honest as Will. In his absence one of his business partners embezzled money from the company, leaving Will and the company nearly bankrupt. His business partner was indicted on three counts of felony in 2009.

Slowly coming out from under the burden of that blow, Will continues to be the troubadour for the eastern hemlock. He is an advisor to filmmaker David Huff, who is putting together a major documentary on the demise of the eastern hemlock titled *The Vanishing Hemlock: A Race against Time.* He is also working with Dr. Robert Van Pelt, Dr. Lee Frelich, and Bob Leverett (cofounder of ENTS and coauthor of *The Sierra Club Guide to the Ancient Forests of the Northeast*) on a textbook titled *Dendromorphometry—The Art and Science of Measuring Trees.*

Will and his ENTS colleagues are pushing at the boundaries of accepted scientific protocol, and they have at least one strong ally in Dr. Van Pelt. Dubbed the "Lorax" by friends, Van Pelt agrees with Will on the proper technique for accurately measuring big trees. (In the Dr. Seuss book called *The Lorax,* the main character is devoted to saving trees despite the relentless march of civilization.) And Van Pelt shares Will's and the ENTS's conviction that finding and measuring big trees is important scientific work.

The addition of volume as a criterion for big tree status in the ENTS's work comes from a February 2004 visit by Van Pelt to join Will climbing and measuring the Middleton Oak in Charleston, South Carolina.

"Two weeks ago I thought that The Senator [a bald cypress in Longwood, Florida] held a supreme position in the East as far as volume was concerned," explains Van Pelt. "I figured that the largest of the Smoky's *Liriodendrons* [tulip trees] would be close to three thousand cubic feet and that the Middleton Oak would come in somewhere around two thousand cubic feet. Boy was I wrong!"

It took Van Pelt, Will, and two other Appalachian Arborist climbers fourteen hours to measure branch segments in the Middleton Oak. Preliminary measurements reveal the following:

Height	67.4 feet
Diameter	10.44 feet
Crown spread	118.0 feet
Wood volume	
Main trunk	970.0 cubic feet
Branches	3,850.0 cubic feet
Total	**4,820.0 cubic feet**

"Wow! Needless to say, I must completely revise my thoughts on eastern trees," wrote Van Pelt in a memo to ENTS members.

Sensitivity to More Than Size

In Will's December 2003 ENTS memo, he wrote eloquently about defoliation and trees becoming "gray ghosts of their former luscious glory." He wrote about looking for answers to nagging questions. And he wrote about species being lost in their prime. These are the words of a man who both studies and loves trees.

Will's climbs have taken him into champions of tulip tree, live oak, loblolly pine, eastern white pine, black cherry, and scarlet oak. On all his climbs, Will educates people about trees and their potential both through the steady work of amassing statistics and through his photography and writing. By photographing trees from the bird's-eye perspective he gains as a climber, he teaches us that trees are far more than just what we see from the ground. As his friend and colleague Robert Leverett, the college instructor dubbed the East's old-growth evangelist, wrote, "Will's trees are veritable hotels in the forest. They harbor so much, much more life and serve so

Will in the tallest known tulip tree, 100 feet above ground. »

many more functions than just storing carbon on the stem. Honestly, forest ecologists need to get up into the canopies of these grand trees and seriously study what's going on."

It is amazing to find so much talent and perspective from a man who has barely crested the forty-year mark. The legacy of his service to the health and survival of trees is already huge. And as this closing piece of writing in his own voice shows, he already harbors the intuition and sensitivity often reserved to artists:

While living in Gatlinburg, TN when working for the Great Smoky Mountains National Park, I was fortunate to have 140–180-year-old eastern hemlocks off the deck of my apartment. Their gnarled crowns and cloaks of lichens were gorgeous, and a regular and vital part of my experience on the property—a respite from the utter hellhole of downtown Gatlinburg.

One evening, while sitting on my deck overlooking the creek and the hemlock grove, I was overcome by a profound, yet unidentified sense of anxiety and unrest as I examined the twisted, gnarled, and flood-sculpted shape of an old, but small hemlock growing in the shade of a large American beech. This little tree had etched into this form the struggles and triumphs of decades of living in the deep shade of the hemlocks and beeches above and a regularly flooding stream below. Some portions of the trunk were straight, others curved, and the branching was at times random and chaotic, reaching for patches of sun both filtered and opened up by fallen neighbors.

I was both saddened by its slow trip to the canopy (what do I know about tree-time, anyway?) and elated that perhaps it was just fine where it was. It was healthy—robust in a way—and the limbs did an elegant twisting dance in the cool winds that flowed down the creek from the 6,600-foot peaks above. I began to think of the tree's life and wonder if it was in any way regretful for its place on the stream, in the dark shade, and on the flood-scoured bank. Was it a patient tree, just waiting for a light gap to tap its reserves and bolt to the sky? Or, had it given up, destined to be an ancient but wispy understory tree with no aspirations to get to the upper canopy?

I continued to study the tree and ponder these thoughts, drawing inspiration from it (I felt it was happy to be where it was) and gaining even more respect for the species and tree timetables in general. Then, just as silent as could be, a huge, dead limb fell out of the large beech above and smashed the smaller tree.

Why was I to witness this moment in time, an instant in the century and a half of the little tree's life?

Part of the reason Will witnessed this dramatic moment is that he is curious, aware, and incredibly alert. He is passing these traits on to his elementary-school-age children—his son has his own climbing harness, and by age two his daughter could identify an eastern hemlock. And he is passing that awareness on to all of us by encouraging us to care about the trees on our own property, intriguing us with the potential of wild trees to grow into giants, participating in National Park Service studies to document attempts to save eastern hemlocks, and challenging us to think about long-term ecological consequences.

Photo Courtesy: Jason Childs

⩘ *Will calling the entomologist of an insecticide company to report on the efficacy of the treatment of the Cheoah Hemlock in the Henry Wright Tract, Highlands, NC*

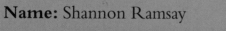

Name: Shannon Ramsay

Occupation: CEO, Trees Forever

Point of Wisdom: Creative alliances of community organizations can provide opportunities for people to develop a land ethic that beautifies their homes, cities, schools, roads, and farms.

CHAPTER 7

The Community Organizer

Shannon Ramsay

A lot of folks don't relate to environmental theory, but if we put a shovel in their hands and give them some instruction about where and how to plant trees, they develop a whole new understanding about the importance of trees in their surroundings.

—SHANNON RAMSAY

IT'S A WARM IOWA SUMMER EVENING. The breeze is blowing softly through the tops of white and red oaks, sugar maples, hickories, and butternuts. The trees make a whispering sound, each broadleaf species distinctive in form, each foundational to the middle section of the North American continent. There is a low rumble of thunder as a distant storm readies itself to bring more rain into the already swollen banks of the Wapsipinicon River. This is the heartland of America, the place where fields of corn and soybeans have replaced the native prairie and where open space still meets meandering, woods-edged streams and rivers.

On the treetop-level deck of her modest, energy-efficient home overlooking the Wapsipinicon, Shannon Ramsay enjoys nature's beauty. This is a rare moment of relaxation for the petite, red-haired community organizer. With the trees as whispering witnesses, conversation shifts to the topic of Shannon's passion, an organization called Trees Forever located in Marion, Iowa. She cofounded this environmental nonprofit with David Krotz in 1989. Intense and intelligent, Shannon is now the CEO and president of Trees Forever and oversees tens of thousands of volunteers, numerous private and public partnership initiatives, and a full-time staff of nineteen. In the first twenty years of its existence, Trees Forever involved over 150,000 volunteers in Iowa's ninety-nine counties and parts of Illinois to plant over 2.8 million trees in towns, along stream banks, and next to roadways. In 2008, when Cedar Rapids, Iowa, experienced a five-hundred-year flood, Trees Forever stepped in to help. And in keeping with its mission to encourage care for the environment, Trees Forever produced the documentary *Spirit of the Trees*, which shares the Native American story of caring for the earth.

Community organizing is Shannon's life work—uniting people in a common cause that offers collective benefit. And since trees beautify communities, provide shade for energy conservation, and contribute to erosion control, it's pretty easy to enroll everyone from government officials to homeowners to rally around the planting of trees.

"From the very beginning, Trees Forever believed in the importance of involving people in what's happening in their communities here in Iowa," Shannon explains. "For many people, Trees Forever is the first step toward implementing a whole new land ethic in their lives. As people get educated about the role of trees, they begin to realize that trees are always dependent on the community of plants around them— just as *they* are dependent on the people around them."

Fading Forestlands

The Iowa Department of Natural Resources estimates that only a third of the forestland that was here when European settlers came still survives. Of the remaining woodlands, 92 percent are privately owned.

Early Years

Shannon's land-based activism integrates and perpetuates the dedication she learned from her parents. Her father, Claude Ramsay, served as president of Mississippi's AFL-CIO from 1959 to 1981. He believed unionism could not succeed in the South until issues such as state reapportionment and the vote for blacks were addressed. He built an alliance between organized labor (whose membership was largely blue-collar whites) and African Americans at a time when both groups were living through huge turbulence and cultural shift. "My father was an activist in Mississippi in the height of civil rights. He was good at standing up for things that were not popular. From him I learned tenacity to values. Perhaps as a balance to all of that, we lived in the country, and my family always appreciated nature. My mother, Mae Helen, was a great nature lover all of her life. At age eighty-four she still regularly visited beautiful natural areas in a wide variety of geographic locations. She has been a huge influence on me. Though they are both gone now, I am conscious that my work continues their legacies.

"I spent a lot of time climbing trees growing up," Shannon remembers. "My favorite tree in Mississippi was probably the live oak, but I especially loved to climb mimosa trees. Being up there, feeling the tree move with the wind, was heaven to me. My indoor passion was reading. One of the strongest memories of my youth was reading every

⤜ *Shannon Ramsay planting a tree in front of the Iowa State Capitol in Des Moines at the rededication of the first Trees Forever project.*

Photo Courtesy: Trees Forever

biography that could possibly help me shape the direction of my life. By fifth grade I had read every biography in my school library."

Shannon attended the University of Mississippi and focused on creative writing and philosophy. After she left Mississippi and eventually moved permanently to Iowa, she went on to manage a number of small companies and work in marketing and promotions. "I was always so values driven that I had trouble matching my need for a livelihood with things I cared about. Without being passionate about something, I couldn't get my heart into the work. So, I did a number of things, including starting my own herbal tea company and managing a wellness company, that were good matches for my values. Trees Forever was the venture that seemed to have all the right things: timing, values, passion, and return to nature."

Shannon and Trees Forever focus on partnerships as key to implementing projects. The project/sponsor list of Trees Forever reads like a list from Who's Who in Iowa and Illinois—the governors, city officials, university professors, Iowa Department of Transportation, Iowa and Illinois Farm Bureaus, public utility officials, agribusinesses, farmers, and more.

Further evidence of Shannon's commitment to networking is her ongoing service as president of the board of directors for the Alliance for Community Trees (ACT). As the only national organization solely focused on the needs of nonprofit and community organizations engaged in urban forest protection, it provides formal training and seminars, grant management, and a collective voice for urban forestry. ACT is the organization that provides help to Trees Forever and three other featured keepers in this book—Andy Lipkis (TreePeople), Corella Payne (TreeKeepers of the Chicago Openlands Project), and Cass Turnbull (PlantAmnesty). Shannon is one of the alliance's founding members. ACT has made a commitment to improve the environment where 86 percent of Americans live: cities, towns, and villages. By 2009, the ACT's national network of members had planted and cared for 14.9 million trees with help from 4.3 million volunteers.

Early Partnerships

One of the first partnerships Shannon and Trees Forever forged was a jointly sponsored program with Alliant Energy called Branching Out. Alliant Energy is a private utility company that serves almost the entire state of Iowa. It shares that territory with MidAmerican Energy and Black Hills Energy and municipal utilities like Waverly Light and Power. "Bob Latham, an executive with Alliant Energy, agreed to be one of our early board members, and that was an incredible gift," explains Shannon. "In

⌃ *Illustration of correct tree placement for energy conservation. Notice compass mark on drawing.*

1989, he helped us join with his company to encourage energy efficiency through tree planting and care. To this day Alliant Energy offers financial support that is matched by local communities who have Trees Forever chapters, discounts from nurseries, in-kind services, and community labor. Trees Forever coordinates these different interests and provides technical and on-site assistance for people who strategically plant trees around their homes to decrease energy usage."

The goal of people inside houses is to maintain a stable interior temperature with minimum use of energy. Correct planting of trees can help achieve this goal. Studying this Trees Forever drawing, notice that conifers (trees that keep needles year-round) are planted on the north or northwest side of homes and businesses to provide shelter from harsh winter winds, reducing energy needed for heating. Deciduous trees (those that lose leaves in autumn) are planted west of west-facing windows and east of east-facing windows to decrease solar gain from morning and late afternoon sun during summer and then increase solar gain in winter. Avoiding tree placement on the south side of homes and buildings allows solar gain in the winter when the sun hovers on the southern horizon. Trees and bushes of any kind are planted around air conditioners and pavement to provide cooling.

In the first fifteen years of operation, the Alliant Energy Branching Out program has saved nearly thirteen million kilowatt-hours of energy as a result of strategic shade tree placement. (A kilowatt-hour is a standard measure of current flow in the same way that miles per hour is a standard measure of speed.) The U.S. Department of Energy figures that a typical American home uses 9,400 kilowatt-hours per year. Even without addressing consumer usage patterns, enough energy has been saved through the program to power 1,383 homes for a year—simply by correctly planting trees!

"Some of the Trees Forever committees in the Alliant Energy program are, of course, more gung ho than others," explains Shannon, "but about 70 percent of them remain active from one year to the next. Active status means they submit proposals for funding." In the first twenty years of the program, over 120,000 volunteers have spent nearly a half million hours planting 1.1 million trees around homes and public buildings. Driving along state or county highways in this breadbasket state, a motorist travels past corn and soybean fields and then enters one forested community after another. There are trees around libraries, city halls, homes, and post offices.

Photo Courtesy: Trees Forever

⌃ *Volunteers from a local Trees Forever chapter at work planting trees along a trail in George, Iowa.*

Even though Iowa is characterized by rural life, the majority of its citizens no longer live on farms. Today most Iowans live in towns or cities. Des Moines, the capital, has over a half million people in the five-county metropolitan area. Many Iowans still feel connected to the farm traditions that have defined their state, though they have lost the skills to effectively participate in caring for the land around them. When people plant trees, it beautifies their community and requires ongoing care to keep the saplings alive. People become caretakers and environmental stewards of their own land and the public spaces in their communities. They get "grounded."

"In 1990, the Iowa state legislature required utilities to become more energy efficient," explains Shannon. "The timing was perfect for Trees Forever. We had been in partnership with Alliant Energy for one year. So, we had researched the correlation between strategic tree placement and energy efficiency. This piece of legislation really launched our programs."

Trees Forever next aligned itself with Black Hills Energy (formerly Aquila), which is a private power company like Alliant Energy that services the entire state. As it does with the Alliant Energy program, Trees Forever assists qualifying Black Hills Energy communities with staff visits, resource materials, and advice on matching funds. Trees Forever has also partnered with local municipal energy companies like Waverly Light and Power (see Chapter 9).

Trees Forever community forestry programs are also funded by local schools, counties, and city governments. The Story City Trees Forever group, founded in 1995, is an active program funded partly by Black Hills Energy and partly by the city. Story City, Iowa, is located in the middle of the state near Ames, the home of Iowa State University. At the 2000 census, there were just over three thousand people living in this small wooded town along the banks of the Skunk River. Many residents commute to Ames for work. The area's Scandinavian heritage is honored annually by a Scandinavian Day celebration.

Mike Jensen has been president of the organization since its beginning. As an employee of the Iowa Department of Transportation and a lifelong resident of Story City, Mike has always enjoyed nature. "Our Trees Forever group has a lot of backing and visibility in Story City. The city council bought a truck for us to water the trees we have planted. Several years ago we lost a lot of trees to a windstorm, so we wrote a grant proposal to both Black Hills Energy and the city and received $4,000 from each of them to replant. In all of our tree planting, we encourage everyone from kids to adults to help out. In years past we have also landscaped the new library, created a seven-and-one-half-acre natural prairie area, and regularly beautified and replenished elderly trees," says Mike. "And I'll tell you this without hesitation: We wouldn't have nearly the number of trees we do in Iowa if it weren't for Shannon Ramsay. She is one great person for the environment."

Roadsides

Once these partnerships were established, like saplings who had taken root and no longer need weekly watering, Shannon shifted her energy to integrate tree planting and environmental beautification on Iowa's roadways. "We have an exceptional

Department of Transportation in Iowa," says Shannon. "They are way out in front of most states in roadside management. About the time we lost the MidAmerican Energy money [1996], we applied for and received federal highway dollars to establish a unique partnership, the Iowa's Living Roadways program."

Designed for Iowan communities with populations under ten thousand people, the program serves as a resource to beautify roadsides or trails with native grasses, wildflowers, trees, and shrubs. Sample projects include community entryway beautification, landscaping recreational trails, highway roadside plantings, roadside rest areas, prairie and savanna establishment, butterfly gardens, and outdoor classrooms. Iowa's Living Roadways program has two components: visioning and projects. The visioning component utilizes Iowa State University interns who are teamed with professional landscape architects to provide planning and design. The project side funds the actual purchase of plants and supplies. And then volunteers get the plants in the ground!

Communities receiving grants have an average population of 3,500 people and receive an average of $7,500. Since 1996, nearly $3 million has been distributed to over three hundred Iowa counties and communities to plant grasses, wildflowers, trees, and shrubs. Roadside maintenance costs have been lowered. Erosion control and wildlife habitat have been enhanced, and people have had the opportunity to beautify their communities.

In 2003, Trees Forever, Iowa State University, and the Iowa Department of Transportation received the Federal Highway Administration's Environmental Excellence Award in the category of Livable Communities, the first such award for the state.

Freedom Tree Memorial and Young People

After September 11, 2001, a Knoxville, Iowa, high school senior, Laura Froyen, became determined to create a local, living memorial to all the rescue workers who had died in service in that catastrophe. With the help of Trees Forever, Alliant Energy, and Laura's youthful idealism, twenty-five beautiful trees were planted. New York firefighter John Mixon of Engine Company Six came for the project's dedication and presented Laura with the helmet he had worn on 9/11 as a thank-you for creating the memorial. In this small central Iowa town, over two thousand community members joined visiting dignitaries to dedicate the installation of flags, decorative flowerbeds, and two rows of *fire* crab apples and *ash* trees, with engraved plaques bearing the names of the memorialized emergency workers. The Knoxville High School science club maintains the site—a key component, as ongoing maintenance teaches generations of local students important lessons in biology, conservation, and history.

"The Knoxville Freedom Trees Memorial is a spectacular example of what can happen around school yards. Schools are often neglected from a landscaping standpoint—too often people put a bush by the front door and blacktop everything else for playground and parking. I love watching parents, teachers, and students get excited about transforming their school yards," says Shannon.

Tree saplings planted by one class become community markers and reference points. A transformed school yard has trees planted around playground equipment and asphalt; it has trees and shrubs outside classroom windows to provide students and teachers with a soothing view. In many communities young parents can show their children trees they themselves planted as students.

In 2005, at Greenwood Elementary School in Des Moines, Jan Berg Kruse, a concerned neighbor to the school, was frustrated by the problem of water runoff and erosion from the hard-surface parking lots on the north side of the school. Instead of getting angry, she got collaborative. Kruse contacted Trees Forever, the Natural Resources Conservation Service, and the Des Moines Public Schools and launched the planting of trees and plants designed to absorb the runoff and create an attractive green space. The whole project took one year and involved all 460 students in the school.

"A lot of folks don't relate to environmental theory," says Shannon. "But if we put a shovel in their hands and give them some instruction about where and how to plant trees, they develop a whole new understanding about the importance of trees in their surroundings. Add to that the responsibility for watering and tending the tree until it is strong enough to stand on its own and you have a complete lesson in environmental stewardship."

Getting a shovel in the hands of young people who are struggling to find a place in their communities has been addressed by Trees Forever's school program, Growing Futures. Partnership is key in this program—between schools or facilities that serve at-risk youth and local experts in the field of natural resources. Youth of racial and economic diversity from Four Oaks of Iowa, the Iowa Juvenile Home in Toledo, and Metro High School in Cedar Rapids have spent time planting, pruning, mulching, and watering trees under the guidance of local architects, extension agents, community college instructors, and industry representatives. By working with professionals, the students are immersed in the basics of care for the natural world and learn about possible environmental careers.

Building on the success and identified needs in the Growing Futures program, Trees Forever has launched an eight-week summer environment work program for youth aged sixteen to twenty-one. Young men and women with perceived barriers to employment are paid for thirty-two hours of work a week while learning about the environment and green careers. Modeled after the highly successful U.S. Forest Service Youth Conservation Corps, this program enables young people to gain solid work experience, develop skills, learn greater appreciation for the natural world, and practice leadership skills.

⩘ *Bear Creek before restoration, a classic "bald" creek. One of the first creek buffer projects of Trees Forever.*

Next Focus—Working with Farmers

The Iowa countryside today consists of field after field of soybeans and corn, dotted with clumps of tall deciduous trees that generally signal homes or farms. Along the horizon, a meandering line of trees outlines the course of a stream or river. What the casual observer does not see are the miles and miles of streams and creeks with no tree protection that abut crops or grazing land. Crops planted to the water's edge promote pesticide and herbicide runoff. Cattle grazed up to the water's edge destroy root structure and bank stabilization. "Bald" creeks and river edges cease to provide habitat for deer, foxes, and pheasants and other birds.

"To address creek-side stress, we needed to partner with the agricultural community," explains Shannon. "So, we went to the Iowa Farm Bureau, and they helped us launch a partnership called Working Watersheds: Buffers and Beyond."

« *The same creek after Trees Forever has restored creek edges to prevent erosion and improve water quality.*

Beginning in 1997, Trees Forever's buffer initiative has established 211 projects and demonstration sites to improve water quality, including the protection of 101 miles of stream bank. At many of these sites, landowners hold field days to demonstrate to their neighbors how buffers of trees and shrubs are working to reduce soil erosion, lessen levels of nitrate and pesticides in the water, and increase habitat. The emphasis on planting some of the buffers in perennial crops that can be harvested has added a new level of creative interest for local farmers. Not only is there a focus on water and soil quality, but also there is a focus on providing an alternative source of income.

Pat and Debra Hayes are an excellent example of farmers utilizing agroforestry practices to reshape their livelihood and update their land practices. Their farm, located in Dubuque County not far from where Iowa, Wisconsin, and Illinois share boundaries along the Mississippi River, includes twelve acres along 425 feet of Whitewater Creek. The Leopold Center of Sustainable Agriculture and Iowa State University are

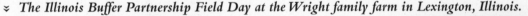

The Illinois Buffer Partnership Field Day at the Wright family farm in Lexington, Illinois.

Photo Courtesy: Trees Forever

funding Trees Forever's efforts to assess market potential for specialty forest products, and the Hayes are happy subjects of the study. Pat and Debra have planted a wide variety of agroforestry crops including plum, native crab apple, nannyberry, oaks, chestnuts, and butternut to sell to nurseries. They have also planted woody decorative florals such as corkscrew, pussy willow, and cardinal red osier dogwood to sell to farmers' markets. In remnant fen and wetland areas, the Hayes have planted native wildflowers such as blue flag iris and New Jersey tea to sell the seeds to native seed dealers. Fruit from the trees is sold to local winemakers, jelly makers, and farmers' markets.

"Our Iowa Buffer Partnership was so successful that farmers across the state line in Illinois began asking for a similar opportunity," says Shannon. In 2001, the Illinois Buffer Partnership was established. Sponsoring partners include the Illinois Council on Best Management Practices, Illinois Corn Grower's Association, Syngenta Crop Protection, Inc., Illinois Pork Producer's Association, Illinois Fertilizer and Chemical Association, Illinois Farm Bureau, and Illinois Soybean Association/Checkoff Board. The partnerships work to improve water quality across Illinois by establishing buffers, wetlands, and other best management practices. In eight years the program has established 146 demonstration sites, creating field days to plant nearly nine hundred thousand plants and shrubs. Landowners choose the conservation practices they wish to implement—stream bank stabilization, stream channel enhancement, constructed wetlands, or plantings around livestock facilities—and the plants they wish to use. The program asks landowners to host a field demonstration day where they share knowledge with interested neighbors and community members.

On June 13, 2008, the Cedar River crested after a huge storm and covered more than ten square miles of Cedar Rapids. This five-hundred-year flood sent water running through

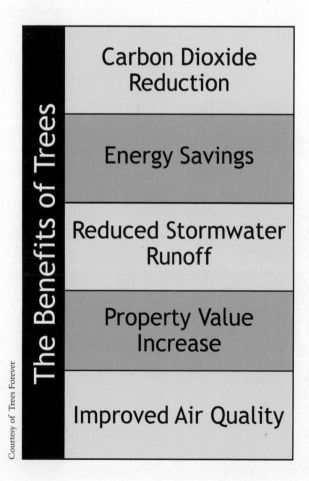

The Benefits of Trees

Carbon Dioxide Reduction

Energy Savings

Reduced Stormwater Runoff

Property Value Increase

Improved Air Quality

Courtesy of Trees Forever

nearly every downtown business and most public buildings. In Cedar Rapids, 944 homes were irreparably damaged, 75 percent of which were low-income housing. A year later residents are still struggling to rebuild. And not unlike the funding issues following the disaster caused by Hurricane Katrina in 2005, federal disaster funding has been slow to arrive and has left entire neighborhoods empty and rotting. Trees Forever stepped forward with an ambitious fund-raising campaign to help. One of the first areas the group focused on was working with Cedar Valley Habitat for Humanity in its efforts to build homes with sustainable landscaping.

All the work of Trees Forever focuses on community development and the five benefits of trees—carbon dioxide reduction, energy savings, reduced storm-water runoff, property value increase, and improved air quality. Shannon loves to quote statistics such as "one acre of trees produces enough oxygen for eighteen people to breathe each day and eliminates as much carbon dioxide from the air as is produced driving a car twenty-six thousand miles." She has extraordinary passion about nearly everything she does—from organizing community gatherings to traveling to farm fields to meeting with boards of directors who may be potential partners in planting trees. But when she talks about the *Spirit of the Trees* project, a special light comes into her eyes. The woman who loves "all of her children" (projects) leans forward to talk about this one.

Spirit of the Trees

"I have always admired the way Native American cultures combine an appreciation and love for the natural world," says Shannon. "Like all states, Iowa originally had a strong settlement of Native Americans who were the first caretakers of the earth here. My professional interest was to share the Native American perspective on the environment."

Shannon believed Trees Forever's mission of planting trees and caring for the environment would be enhanced by sharing the Native American story of caring for the earth. Her first and most important alliance was with Maria Pearson, considered the mother of the movement to return native artifacts to native people.

Iowa was the first state in the nation to pass a law requiring the protection of Native American burial grounds and sacred items. When Iowa governor Robert Ray signed the bill into law in 1976, Maria was there. One hundred percent Yankton Sioux, Maria was married to an Iowa Department of Transportation road engineer. She personally shepherded the passage of the bill and launched a national and international effort to protect ancestral remains.

"I first met Maria Pearson in 1995 at the Mandarin Chinese Restaurant in Ames, Iowa. Later I would discover that the Mandarin was in many ways Maria's second office. The family that owned it adored her—treating her like royalty." As the women came to know each other, Maria taught Shannon to take time in the natural world for ceremonies of gratitude and appreciation, no matter how pressing her time schedule.

Together, Shannon and Maria launched the idea of a documentary film on Native Americans' connections with trees. "We wanted to do a piece that featured Native Americans, not scholars talking about Native Americans. I knew that Maria would be a highly regarded ambassador for the film project, opening doors to tribes and nations across the United States. We found a nationally recognized documentary filmmaker, Catherine Busch-Johnston of California, who was known for her compassionate, quality work. My good friend and colleague Janette Monear with Tree Trust of Minnesota finished off our team as the educational consultant. We submitted a proposal to the U.S.D.A. Forest Service and received funding in July of 1997 to begin the *Spirit of the Trees* educational documentary."

In the next two-and-a-half years, the team made six major trips from Hawaii across the continental United States and included footage of over one hundred pieces of Native art as part of the final six-part series *Spirit of the Trees*. They interviewed over one hundred participants from forty different tribes and nations. "The trips were a crash course on traditional Native American ways, and along with other team members, I regularly got failing grades. Maria instructed us how to treat elders, how to work with tribes, and how to be respectful. She taught us with humor and sternness. I discovered that one of the greatest Native American gifts is humor.

"Most of the Native Americans we met on our journeys had been early 'warriors' with Maria in the movement to protect and repatriate ancestral remains. For example, Maria was in New York City with the Native Hawaiians when the bones of their ancestors were returned by the Smithsonian."

One of the many stories Shannon tells about the learning process of making the film is about when they traveled to the southeastern United States to interview the Seminoles. "We had made numerous calls ahead but had no official clearance to travel the reservation and conduct interviews. We managed to reserve some cabins at the Seminole Big Cypress Reservation in the Everglades and just flew down. After breakfast the first morning, we were driving our rented van when a large SUV pulled up to the stop sign near us. Maria quickly ordered us to stop and hopped out to wave down the SUV. She went over and warmly shook hands with the driver, Chief James

Billie, whom she knew well. He then personally took us over to the home of a traditional canoe maker. From then on, every door was open to us."

Maria died before the documentary film won the prestigious Videographer Award of Excellence, two Telly Awards, and an Intermedia-Globe Diploma Award. Her death was a great loss to many, including Shannon, and the six-part educational video series is part of her legacy. Exploring the link between trees and forests and Native American culture, *Spirit of the Trees* weaves native voices, art, and music from tribes from the central United States, the northwest, the northeast, Hawaii, the southwest, and the southeast. High-quality copies of all footage of interviews were returned to the participating tribe or tribal members, along with a copy of the entire series. The documentary can be purchased from the Canadian Film Board via the link on the Trees Forever Web site. The film is a powerful tribute to "the standing ones," the Native American term for trees, and to many tribal elders.

Flagships of the Environment

"Trees are flagships of the environment," says Shannon, back on her deck. "We relate well to them because they are large and charismatic. And they are so much more . . . home to many creatures, providers of shade and protection from winter winds, purifiers of the air, and suppliers of lumber, furniture, and food. Trees are simply one of the most valuable natural resources on the planet."

Shannon works hard to carry these lessons of trees to the people of Iowa, Illinois, and the nation. She uses her considerable networking skills to gather people of all ages, persuasions, and skills to work together for the common cause of creating a healthier, more beautiful place to live. Like the saplings she has inspired people to plant all over the states of Iowa and Illinois, her contributions keep growing.

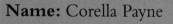

Photo Courtesy: Ann Linnea

Name: Corella Payne

Occupation: Public health researcher, gardener

Point of Interest: By spending time volunteering to care for trees in public parks, city dwellers provide an important service to their community and to their own souls.

CHAPTER 8

The Weekend TreeKeeper

Corella Payne

People can't be surrounded by granite, cement, and asphalt all day. It's not good for the spirit. Find something green and natural to rest your eyes on.

—CORELLA PAYNE

MAPLE TREES ARE BEGINNING the slow turn to crimson. Temperatures are cool—it's a wool cap and gloves kind of day. Birch, hickory, and sycamore trees are already golden in the great city of Chicago. Corella Payne, an African American woman in her late forties, is spending her day off from work pruning, raking leaves, and spreading mulch as a TreeKeepers volunteer. She rests a moment and stares pensively over the landscaped grounds of a city park.

"I have a stressful job during the week. I get tired of talking and listening. People tire me out, then nature soothes my soul," explains

Corella. "I could spend my Saturdays at a mall buying stuff that I don't need, but what's the sense in that? When I finish tree keeping, I've had a workout, I feel like I've done something good for the planet and useful for my community, and I haven't spent a dime. It's so much more relaxing than running up a credit card."

Corella's recent job as a research assistant on an asthma study project took her into the homes of many lower-income African Americans, Latinos, and others suffering from asthma. She gathered physical and sociological data from both asthma patients and their caregivers, including drawing blood samples. And though she may not have known it at the time, her work and her tree tending are connected. The *Journal of Epidemiology and Community Health* (May 2008) published research that supports Corella's belief that time outdoors and time with trees translates into improved health. Specifically the study showed the prevalence of asthma in inner-city children could be correlated to the number of trees on the blocks where they lived: more trees equaled fewer cases of asthma.

Corella's job required flexible hours, including weekends, and a great deal of sensitivity as she dealt with all of the complexity of lives being lived at the economic edge and complicated by chronic illness. She is quite aware that with two master's degrees, she is overqualified and underpaid for the work that she is performing. "It's okay for now," she says. "I'm just grateful to have even a short-term job with health insurance.

"A few years back I found myself getting crabby, irritable, and restless," explains Corella. "People were either a disappointment, a frustration, or a headache, much like life can be on a regular basis. I decided to find some kind of recreation that had more to do with nature than with people. One day I went over to Lincoln Park Zoo to see about being a gardener or doing something that would allow me to spend time outdoors."

Corella's irritability is not unusual in high-density, urban environments characterized by pavement, skyscrapers, indoor malls, constant traffic, and crowds of people at every step out of the house. In most of human history in the past 165,000 years, and most of human experience until the twentieth century, people's lives were grounded in the opposite of these conditions: living with expansive nature in rural environments, living in and with the elements of weather and cycles of the seasons, being educated about the local plants and animals, and living in nomadic small villages or tribal groups.

It seems as soon as we turned our backs on this style of living in Euro-American cultures, scientists began studying the impact of industrialization and urbanization

Chicago's nearly five million trees are contained in an extensive system of parks both within Chicago and Cook County. »

on the human organism. Writing in the mid-twentieth century, Erich Fromm and Edward O. Wilson developed and explored what they called the "biophilia hypothesis," the inherited and instinctual bond between human beings and the living system that sustains us. "Biophilia" means "the love of life, or the love of living systems." Corella's impulse to restore herself by returning to nature and reconnecting to the living system around her is brilliant common sense.

In 1996, while serving as a volunteer gardener at the Lincoln Park Zoo, Corella saw a sign advertising a seven-week training to become a TreeKeeper. "It looked interesting to me—kind of a technical training about trees—species, pruning, mulching, and so on. I signed myself up and have been working with the program ever since."

TreeKeepers is one small program of the Chicago Openlands Project. Volunteers pay $75.00 for the training that covers such topics as benefits of urban trees, tree

Photo Courtesy: Ann Linnea

folktales and myths, tree physiology, tree species identification, common tree disease and insect damage, and hands-on pruning, planting, and mulching. Local professionals such as Dr. George Ware, retired director of Morton Arboretum and a world-renowned expert on American elms, teach classes.

The Morton Arboretum is a Chicago-area treasure located on the west side of Cook County in Lisle, Illinois. With the intended purpose of collecting and studying trees and shrubs from around the world, the arboretum has an extensive array of ongoing educational programs for children and adults, including a cooperative botany degree program with regional colleges and universities. The arboretum offers specialty displays like the Schulenberg Prairie and the Children's Garden, miles of hiking paths, and exquisite landscaping in every quadrant. Maintaining the Morton Arboretum requires an intense amount of work and consequently provides golden opportunities for urban dwellers like Corella to spend time outdoors volunteering and learning something new about nature.

"The training is a big commitment," explains coordinator Jim De Horn. "Imagine finding seven consecutive Saturdays to devote to any training. Close to half our TreeKeepers don't get through the course and exam the first time. But when they do emerge, they are ready to help us care for our urban forests. After twelve years of trainings, we have over 650 TreeKeepers doing their best to care for some of Chicago's five million trees."

All of those trees are contained in an extensive system of parks both within Chicago and in Cook County. The Chicago Park district manages 7,600 acres of parkland, 570 parks, 24 miles of lakefront, 22 prairies and grasslands, and 11 savannas and woodlands. The Forest Preserve District of Cook County covers an additional 67,800 acres, 11 percent of the county's landmass, making the greater Chicago area a haven for people seeking time in nature. Corella's city prides itself on its parks and making them available to citizens of all economic backgrounds. And it was Grant Park that was the only space in Chicago big enough to gather and hold one of the largest and most diverse crowds in U.S. history—250,000 people on election night 2008.

Early History

Like many of the tens of thousands of volunteers that form the backbone of non-profit tree-planting and maintenance programs across North America, Corella is not a professional gardener or someone previously affiliated with outdoor activities. She is simply a good soul who has chosen to spend some of her volunteer time and energy planting and caring for trees. Her life as a public health worker doesn't give many signals that she would even enjoy spending time with trees, but a closer look at her early years shows someone who spent time in nature.

One of five daughters, Corella became active as a Girl Scout counselor. Scouting got her outside in a helping capacity. After leaving home, she went to live in, study in, or simply pass through diverse living environments in Wyoming, Colorado, California, New Mexico, Texas, Washington, and Alaska. Living in different bioregions opened her eyes to the diversity and majesty of the natural world. Although her formal education took her out of the woods and into people-oriented fields to earn master's degrees in both education and public health, she always retained a certain level of comfort and connection to being outdoors.

Comfort outdoors has been the hallmark of childhood until the last generation or two. Children need nature. Famed naturalist Rachel Carson wrote in *The Sense of Wonder*, "A child's world is fresh and new and beautiful, full of wonder and excitement. If I had influence with the good fairy who is supposed to preside over the christening of all children, I should ask that her gift to each child in the world be a sense of wonder so indestructible that it would last throughout life, as an unfailing antidote against boredom and disenchantments of later years, the sterile preoccupation with things that are artificial, the alienation from the sources of our strength." Corella articulates well the struggles she has with the disenchantments of midlife, and by choosing to be a TreeKeeper, she demonstrates the wisdom to keep reattaching herself to her sense of wonder about the natural world.

"Tree keeping brings me a lot of special moments," she says. "One day I was pulling weeds while looking at some birds on the ground near me. My eyes traveled upward as I heard knocking. I saw two redheaded woodpeckers in a nearby tree. My heart just soared. You can't buy a moment like that at the mall, but you can savor it and carry it with you through the day."

Journalist Richard Louv, author of *Last Child in the Woods: Saving Our Children from Nature-Deficit Disorder* (2005, Algonquin Books) spent nearly a decade interviewing children, parents, teachers, and researchers and concluded that children are becoming increasingly alienated from the natural world, with grave consequences to their physical, mental, and spiritual health. He attributed children's decreasing time outdoors to social environmental factors—dwindling wild spaces for children to play safely, parental fear of "stranger danger," and the seductive availability of technology such as computers, cell phones, Game Boys, and video games. Corella grew up in a world more like the one Rachel Carson describes than the one documented by Richard Louv. However, as an adult living in one of the largest American cities, she has found herself working with and surrounded by people who don't think about going for walks in the city's parks. It takes a lot of energy to head out the door when no one else thinks that is a good idea.

After completing her second master's degree, Corella joined the Peace Corps—something she had always wanted to do. "The Peace Corps satisfied two goals for me: going somewhere for an extended period of time and doing something to make a difference. It wasn't just about me," she explains. Having had several years of Spanish at university, Corella felt comfortable with the language and asked for a region in the Americas. She was selected to serve in Ecuador.

She arrived in Ecuador with a substantial background in the public health arena of HIV disease. "At the time there was a rather limited awareness of HIV disease in Latin America and in Ecuador," explains Corella. "Its eleven million people were very Catholic, very traditional, and fairly conservative.

"Given prevailing attitudes and values, I simply could not see how Ecuadorian women would step forward and seek help from public health officials on HIV/AIDS issues." But Corella is a true public health activist. She was convinced that there were women and children with HIV/AIDS in that country despite official claims to the contrary by Ecuador's public health specialists. During her 1991–95 Peace Corps placement, she made a number of connections with women dealing with HIV/AIDS themselves or in their children. When she returned to the United States, she continued to maintain those Latin American connections, including making self-paid trips back to Ecuador in the fall of 2000, the fall of 2003, and the spring of 2004. In July 2005, because of her specialization, she returned to Ecuador for a six-month second placement.

Corella does not forget the women and children of Ecuador. She relies on her understanding of nature's cycles to provide a sense that engrained cultural mores are capable of shifting. "Things take time," she tells herself. "Trees don't grow tall overnight. Change can take years." The incredible stamina and courage required to work so closely and openheartedly with the complicated plight of people takes a toll on Corella's caring heart. Too much sometimes. So, perhaps that's why she feels called to turn some of her compassion toward hands-in-the-dirt tending of Chicago's natural areas.

Transformative Moment

"My mother always used to say to me that I needed to learn how to come apart from the world, to learn to wear the cares of the world like a loose garment. I used to wonder what that meant. Now as I get older, I have learned that means to separate my joy of life from the issues of the human world such as sexism, racism, and violence. When you die, the world's, troubles are all there anyway. So, you let it go." What remains, is nature.

Corella feels better after caring for trees. She finds great comfort being in the presence of such large, gentle beings. "Trees don't wage war, and they don't kill people. Nature is just nature. I take relief in that."

A short while after Corella finished her TreeKeeper certification, she found herself staring at a young maple tree that had a beer can stuffed in a lower branch. "I spoke

≳ *Seeing a beer can stuffed in a young maple tree was a transformative moment for Corella.*

out loud, to no one in particular, 'How could anyone put a beer can in a tree? Trees are living things, too!" I wasn't being spiritual or mystical. In fact, I've been fighting with God most days of my life. But I was so sorry to see that can there. It really triggered something in me.

"I don't know if I can describe it," says Corella. "I didn't find the muse or anything like that. I didn't find God or whatever that is. I just saw the tree for what it was—calm, good, completely innocent, just being a tree and minding its own business. And someone had had the nerve to defile it." From that day forward, Corella became an activist about caring for trees.

On weekdays or a weekend that Corella does not have to work, she volunteers as a TreeKeeper in the city parks. Some days she pulls thistles, prunes hedges, or cuts back sucker shoots at the bases of trunks. Other days she spreads mulch to retain moisture and keep down weeds. Always she picks up trash people have carelessly tossed aside.

"I turn my caring energy to trees because trees don't disappoint me. They don't cause hurt, grief, or pain. Trees are healing for me because they don't have anything to do with humans. They are timeless. They are grounded and forever and are here long after we're no longer here. For me, caring for trees helps me keep life in perspective and keeps me from feeling overwhelmed or swallowed up by the burdens and challenges of living. I find comfort in nature, the seasons, and the cycle of the universe. Like my mother said, when I come apart from the world, I am refreshed and renewed and can come back and try living again."

Mentor

During her TreeKeeper training, Corella had "the great and good fortune" to meet Ms. Mary Peery. Though she was seventy-nine at the time, Ms. Mary still stood six feet tall. Corella loves to describe Ms. Mary's fight with Chicago's city hall to get an abandoned, drug-infested building on her block demolished. "I'm not sure how she pulled that off. Ms. Mary is a sweetheart, but I would not even dream of crossing her. She gives off an air that says, 'You know I'm a nice person, but please do not, *do not ever* take my kindness for weakness.' Anyway, that building came down, and Ms. Mary got a huge community garden planted there.

"Ms. Mary calls that garden Paradise, her heaven right here on earth. She gets up real early in the morning and tends the tomatoes or the beets or the carrots. She checks on the flowers and fusses with whatever is growing at the time. If I had my choice between going to church and working outdoors, I would go to Paradise just to be in the garden with Ms. Mary any day.

"One year I had a lousy-paying job, and I had a hard time making ends meet. My attitude gave new meaning to the word *nasty*. I was always hard up for money and struggled for months just to make my rent. I was not a nice person to be around. I could barely stand my own company. When I could scrounge up the carfare from the nickels and dimes in my coffee cans, I would make the long trip out to west Chicago to hang out in Paradise.

"Raking, weeding, pruning, and clipping, I would work off my anger and frustration about being so broke. Gardening was a physically hard, healing workout. Afterward, Ms. Mary would give me some beets and some tomatoes and other vegetables to take home. My poverty wasn't solved, but my heart and soul were nourished enough for me to keep going. Ms. Mary understood what I had gained in ways I couldn't. Before I left, she would always give me a hug and say, 'It's going to be okay.'"

For Corella, Ms. Mary is a treasured mentor. "She is truly living history, and her garden is a testament to how to work in nature."

Breast Cancer Walk

Corella's volunteerism keeps getting her into nature in ways that soothe her committed soul. In June 2004, she completed the Avon two-day, thirty-nine-mile walk for breast cancer. Two months later in August, she also completed the Susan G. Komen three-day, sixty-mile breast cancer walk from Wisconsin to Chicago. She participates in the walks to remember the daughter of a former executive director, the mother of her Ecuadorian Peace Corps counterpart, and Ms. Mary, who is a double breast cancer survivor.

Corella likes to walk and understands how it benefits her whether the "cure" she is walking for is cancer research or personal contentment. "I live right by Lake Michigan, I've got a good pair of gym shoes, and walking is a heck of a lot cheaper than indoor fitness places like Bally Fitness or Women's Workout World. Walking always helps me get heavyhearted stuff out of my system. And it's a great way to be outdoors.

"That summer of 2004, when I was training for the cancer walk, I would be stomping down a park path lost in my own internal rant. That year I was thinking about the war in Iraq and how the president and a lot of mostly white men in power thoroughly piss me off. All of a sudden, I heard a flock of Canada geese and turned to see them flying over Lake Michigan. The lake shimmered in at least five different shades of green and blue. The geese were flying overhead in a flapping 'V' formation. And all I remembered thinking was, 'Never mind the world and destructive people. No human with a PhD or MD degree or any other made-up human title can explain

Photo Courtesy: Jeanne Petrick

⌃ *Corella finds peace walking alongside Lake Michigan.*

why geese fly that way. We can make a machine fly, but we can't fly. We don't make rainbows. There's not a human power on earth that can make the moon rise or the sun set. These are things bigger than us, beyond us. Things over which we have absolutely no control. And it gave me comfort. Nature is beyond us and will go its own way. Nature will do what it does, regardless of us. *Thank goodness.*"

Advice from Corella

When asked what city dwellers who don't have a yard can do to connect with trees, Corella responds, "Find a tree-tending program like ours in your city. Or just go find a park, walk in it, and notice what needs doing. Is there trash to pick up? Do the trees have a bunch of weeds or other plant life choking their growth? Pull them up and put the weeds in a garbage can. Don't make another mess while trying to be a tree keeper.

"People can't be surrounded by granite, cement, and asphalt all day. It's not good for the spirit. Find something green and natural to rest your eyes on. Go take a walk outside and try listening to birds chirp. The first time I realized I was listening for a certain bird with a unique song, I followed a bright red cardinal. My heart was full for a long moment after that."

Corella's commitment to volunteerism helps keep her optimistic even when she's frustrated by low-paying work. "Some of the stuff we do in life should have nothing to do with money. Being a TreeKeeper for three to four hours a week keeps me from spending money or even thinking about the economy of my life. I learn how to give back and connect with something much larger than myself, like the whole cycle of life. Trees give me something to hold on to in an ever-shifting and changing world."

In the winter, the great hardwood forests of Chicago lose their leaves and rest. After the last leaves are raked, TreeKeepers like Corella set aside their gardening gloves, and for a few months there is no tree keeping to be done. But Corella doesn't hibernate. She dons her warm clothes and takes lots of walks on frozen or snow-covered ground beneath the naked silhouettes of the city's trees.

"I just try to take wonder in it all, to let my soul rejoice in the resilience of nature. I know the leaves will come back. Winter is nice because there aren't so many people around, but I always look forward to spring and grounding myself once again in the caretaking of trees."

≽ *Corella's advice: "People can't be surrounded by granite, cement, and asphalt all day. It's not good for the spirit. Find something green and natural to rest your eyes on."*

Photo Courtesy: Ann Linnea

Name: Glenn Cannon

Occupation: Businessman

Point of Interest: Trees provide us with a triple whammy for our investment. They beautify our cities, they increase energy efficiency, and they reduce greenhouse gases.

CHAPTER 9

The Businessman Who Loves Trees

Glenn Cannon

As public utilities, we have a monopoly. Along with those rights come responsibilities.

I believe we have a moral obligation to be good stewards of the resources we work with.

—GLENN CANNON

SOMETIMES THE LIFE WORK of a keeper of the trees happens inside, at a desk that looks out on a view of trees. Glenn Cannon came to Waverly, Iowa, in 1990 to become general manager for a small, municipally owned power company and helped turn it into a national model for integrating trees into the corporate mission for energy efficiency. Now in "retirement," Glenn has been freed from his desk to travel nationally and internationally to share his understanding of the role of trees and energy efficiency in helping businesses and communities turn toward deeper and deeper commitment to a green future. One career has led to the next career; one story has led to the next story.

Waverly, Iowa

⌃ *Waverly Light and Power is locally owned and operated.*

Waverly is a community of nine thousand people, a forested oasis in the midst of farm fields that stretch to the horizon in all directions. The flourishing forest is partly attributed to the community forestry program initiated and funded by Waverly Light and Power when Glenn was hired. He is quick to share credit and feels strongly that he was simply carrying on a long-standing community tradition to protect trees. Located by the banks of Iowa's meandering Cedar River and surrounded by tall oaks and maples, the office of Waverly Light and Power is a recently built, two-story brick structure with management offices in front, generators and service trucks in the back. In one of the front offices with a window view of green lawns and pleasant landscaped shrubs, general manager Glenn Cannon worked for seventeen years.

Out of the office on a summer day wearing a business suit, Glenn never seemed to be bothered by temperatures pushing 90°F and the humidity hovering at 95 percent. He often walked along the shaded path from his office to city hall. Arriving at the Waverly Civic Center, he loved to stand next to the old burr oak and share its story. "Citizens did not want to see this tree taken down when city hall needed to expand in 1991. They led quite a fight to save it. I had been in town only a year. The activism around this tree was inspirational and let me know that the community, as well as the company, was open to taking up issues of conservation."

Waverly Light and Power is a relative rarity in the United States today. Most communities receive their electricity from giant corporate utilities. People flip the switch, and the lights come on—or not—and they have no idea where their electricity is coming from. But in Waverly, about a century ago civic leaders of this small farming community decided to control their future electrical service by retaining ownership of the river's dam and purchasing the electric plant for $13,500 from a private enterprise. The city did not have that much money and couldn't raise enough through issuing bonds. However, four newspapermen pooled their resources and offered the city an $8,000 loan with an option to buy out their shares at a later date.

The burr oak saved by citizens of Waverly, Iowa. »

Today, the city owns a utility with assets of over $50 million. The profits of this utility have helped to finance a new city hall, a library, and two hospital expansions. Glenn's management helped keep the company in the black and the community in the green. Trees are everywhere around Waverly, and most of the newer plantings have been placed strategically for energy efficiency.

Early History

Glenn's environmental conscience started at a young age. "My grandpa, E. A. Woods, my mother's father, was a career forest ranger in Montana, and I spent some of my first years learning from him about vastness and the importance of western wilderness. His dedication to the forest and his understanding of both the biology of trees and the ecology that sustained them made a huge impression on me. My father also spent a lot of time with me in the woods, hunting, fishing, and rock hunting. When I look back on the trips with my father and grandpa, I was able to go out and enjoy forests that were bigger than some of the countries I would later live in. I had a vision of what a lot of trees looked like, and I bonded to that beauty."

Since Glenn's father was in the military, and Glenn himself spent six years in the Navy, he has lived in Spain, Okinawa, Japan, and numerous U.S. states. His father's hobby that traveled with them through all these settings was woodworking. "So, when I was indoors and far from Montana, my dad taught me how to make things out of wood. That helped me appreciate the aesthetic of trees from another standpoint. Out of my whole life, and all the things that come and go, the possession that I'm the most proud of is a solid cherry chest patterned by the great eighteenth-century New England woodworker John Goddard. It is absolutely beautiful."

Once out of the Navy, Glenn took his college degree in economics from Clemson University and his leadership skills and in 1977 went to work for Santee Cooper, a state-owned utility in the lowlands of South Carolina. In this first job, he worked with customers to help inform them about what they were buying. "This put me on the other side of the meter—that is, focusing totally on the customer. It was my job to help people understand how their homes used energy and to find ways they could reduce their energy bills. For example, I would explain to them that a new refrigerator uses half the power of an old refrigerator and still does the same job. This helped reduce the customer's electrical bills and, of course, enabled the company to save power. This is commonplace now—with the Energy Star products—but back then it took a lot of education." He liked the work, discovered that most people want to save both energy and money, and stayed in the field to build his career.

In 1990, Waverly Light and Power asked Glenn to serve as its general manager. The company's specific interest was to find more ways to focus on energy conservation. "When I found out that they were serious about conservation, my wife and I moved to Iowa and put down roots."

Waverly Light and Power

In his first year in Waverly, Glenn continued the work he had done in South Carolina by launching an aggressive energy-efficiency and customer-awareness campaign to get people to purchase energy-efficient appliances. Within three years, Waverly Light and Power had saved $235,800 as a result of these efforts. In the late 1980s, the utility had commissioned a study that showed Waverly's energy demand was growing at 4.2 percent, almost twice the national average. As a locally sourced supplier, Waverly Light and Power knew the increasing demand was not sustainable, which is why it hired someone with an environmental consciousness.

A tree hugger in a business suit, with a clean-cut look, Glenn knew he had to balance a strong environmental message with a commitment to the bottom line of profit, so he moved carefully to make changes that would shift Waverly into a town that was reducing its energy demand. In his first year of work, he attended a state association meeting for municipal power companies. It was there he heard Shannon Ramsay of Trees Forever speak about establishing community forestry programs. "I knew Iowa had lost a lot of tree cover to farming. I could see Shannon's nonprofit organization had good information about community forestry and its contribution to energy saving and city beautification, so I signed Waverly Light and Power up as a sponsor. We've been collaborating ever since."

Since 1990, Waverly Light and Power has pledged over $10,000 per year to Trees Forever, which has in turn channeled that money into the local Trees Forever chapter and helped it create a successful tree-planting program. According to the Waverly Light and Power Web site, "By planting a tree properly you could save close to fifty percent or more on the air conditioning portion of your bill." This is done by encouraging homeowners to plant deciduous trees on the south and west sides of homes for shade in the heat of summer. A 50 percent savings sounds like a lot, but Trees Forever has done much research to show the bottom line of profit that can come from the simple act of correctly planting the right species of tree in the proper place.

In the first couple of years in his job, Glenn had launched two major programs to help his company decrease energy usage: a rebate program for energy efficiency and an alliance with Trees Forever to create a tree-planting program. The step he

⌃ *In 1993, Waverly Light and Power became the first public power system in the Midwest to own and operate wind generators. "Today Iowa is second only to Texas in wind production," Glenn explains, "But back then people at the state level just laughed at us."*

was about to take put him way out on the edge of how Iowans were thinking about power.

In 1993, Waverly Light and Power became the first public power system in the Midwest to own and operate wind generators. As a member of the American Public Power Association, Glenn had been learning about wind power and the availability of funds for initial investments. "Today Iowa is second only to Texas in wind production," Glenn explains. "But back then people at the state level just laughed at us. I couldn't even imagine that a decade later the Iowa Economic Development Agency would be printing a map showing all the Iowa wind farms. However, I'm glad I took the risk when I did because it put us in a leadership position regarding energy efficiency and conservation. Consumer change is essential, and so is change at the business level."

Between 1999 and 2005, several wind turbines were added to the Waverly Light and Power energy production. In 2009, the Cannon 1 wind turbine was built and named after Glenn. Along with the Skeets 4 Tower installed in 2001, the Cannon 1 produces enough energy to serve 782 homes in Waverly. These two-hundred-foot towers are bolted into the ground at a depth of thirty feet so they can withstand even the strongest winds. Each year, wind generators get more capable of producing more energy with less wind by improving the technology and the control systems.

Since 1993, Waverly Light and Power customers have received more than $6.1 million in savings because of energy-efficiency programs that include proper tree planting, energy-efficiency rebates, wind power, and system upgrades. Commitment to energy efficiency has earned Waverly Light and Power numerous national and state awards, including the Iowa Energy Leadership Award.

American Public Power Association

In 2003–04, Glenn's influence became more national in scope when he served as board chair for the American Public Power Association (APPA). Created in 1940 as a nonprofit, nonpartisan organization, APPA serves two thousand community-owned electric utilities representing forty-five million people in cities as small as Waverly and as large as Seattle and Los Angeles.

By the year of his term (2003–04), global climate change was beginning to come to general attention. It was obvious that trees had an important role to play in removing carbon dioxide, one of the principal causes of global climate change, from the atmosphere. Glenn coined the phrase "triple whammy" to talk about trees as contributors to increased energy efficiency through proper planting and beautifiers of cities and as reducers of greenhouse gases. "'Triple whammy' is my catchy phrase—it catches the bottom-line thinkers," Glenn says with a smile. "You've got to explain trees in terms of value, not aesthetics. I love a forest because it's a forest, but many people working in public utilities haven't had that experience. They didn't have my grandpa. So I talk to them in terms they understand."

During his term as board chair of APPA, Glenn placed emphasis on strengthening one of APPA's programs: Tree Power. The program encourages member public power utilities to plant at least one tree for every customer they serve. As of 2009, over 267 public utility companies from all over the United States serving 11.8 million customers (an increase of 3 million between 2006 and 2009) are involved with Tree Power. This is a commitment to planting 11.8 million trees!

Glenn learned many things working with APPA. When asked for an example, Glenn responds, "Gainesville Regional Utilities in Florida

Triple Whammy of Trees

1. Increased energy efficiency
2. Reduced greenhouse gases
3. Raised property value in beautification

® « *Logo for the Tree Power program of the American Public Power Association.*

conducted a study that showed areas in their service territory with significant tree cover had much lower incidences of lightning strikes on their lines and infrastructure compared to areas with little tree cover. As a result, lower incidences of power outages and interruptions occurred. This represents significant cost savings to them and their customers."

A Tree's Monetary Value

During Glenn's time as APPA board chair, the "Tree Benefits Calculator" was developed by member utility SMUD (Sacramento, CA, Municipal Utility District). The Tree Benefits Calculator allows utilities to calculate the energy savings and carbon sequestration resulting from planting trees in urban and suburban areas. The utility managers figured out how to quantify the value of one shade tree.

"There is a savings to the customer, to the utility, and to the environment for planting trees in the right place," Glenn explains enthusiastically. "And for utilities that haven't really gotten started on this yet, here is a tool that enables them to calculate the benefits before they even start. Furthermore, utilities and the public can figure out how to calculate benefits and explain these benefits to their customers and stakeholders."

The calculator is a tremendous, free tool available to anyone interested in planting trees to get the most benefit from them. By connecting to www.APPAnet.org/treeben, people can obtain a detailed explanation of the formula and get answers to commonly asked questions ranging from "What if my tree species isn't listed in the index?"

The following is the information one needs to use the Web-based tool:
- the tree species,
- the direction the tree faces,
- the distance between the tree and the building being shaded, and
- the age of the tree from the tree-planting date.

to "In figuring my tree's age, what if my tree was planted from a five-gallon pot instead of a one-gallon pot?"

The Tree Benefits Calculator can also be used to solve complex problems and resolve controversial issues surrounding trees. For example, when Waverly citizens rallied around the burr oak in front of city hall, one party wanted to preserve it, and another wanted to remove it. If that controversy happened now, the Tree Benefits Calculator could provide an impartial, unemotional, and accurate piece of information to aid in the decision-making process. Or say the developer of a new neighborhood wants to market energy efficiency and sustainability as part of the package to new buyers. The Tree Benefits Calculator can help him preserve existing trees and plant new ones throughout the development in ways that benefit home buyers and set guidelines that offer triple whammy enhancement of property values.

Trees as Reducers of Global Climate Change

In different parts of the world, trees not only look different but also serve people and ecosystems in different ways. For Andy Lipkis and his TreePeople organization (Chapter 1), trees are needed to help capture and store precious rainfall for a desert city's use. In the center of North America where the climate cycles range from hot, humid summers to cold, dry winters, trees are needed to help minimize heating and cooling requirements of buildings. However, in all parts of the world, trees are increasingly understood as an essential response to global climate change (also referred to as global warming).

Since the turn of the century, global climate change has gone from a concept that scientists were trying to warn us about to an international mainstream concern mentioned in media and political speeches and considered at global policy and treaty meetings. Global climate change is being caused by an abnormal buildup of atmospheric greenhouse gases, largely carbon dioxide (CO_2) but also methane, nitrous oxide, and chlorofluorocarbons that hold in the earth's heat and increase global temperatures. As Glenn sees it, global climate

Photo Courtesy: Waverly Light and Power

Glenn speaking at a local college about the importance of a locally owned and controlled utility focusing on power production and environmental stewardship. »

change should be a primary concern of all power companies. "Approximately one-third of all CO_2 emissions due to human activity come from burning fossil fuels to generate electricity. One-third of a global issue is a sizeable chunk of problem utilities could be working to resolve. The larger the population in a service area, the more electricity is required to keep people comfortable, functional, and employed. The more electricity produced using fossil fuels, such as coal and natural gas, the higher the production of CO_2 as an emission by-product," explains Glenn. In the "Business and Ethical Case for Energy Efficiency" presentations he is doing in retirement (for Plains Justice, a public interest law center), Glenn reiterates the moral obligations of environmental stewardship and education for energy efficiency.

"On the one hand, it's our job as utility managers to meet demand in a cost-effective manner. However, if the role of a locally-owned and locally-controlled utility is expanded to encompass power production *and* environmental stewardship,

Photo Courtesy: Waverly Light and Power

⌃ *When campaigning in Iowa during the 2008 election, President Barack Obama kept hearing about the small city that was making big contributions to energy efficiency. He traveled to Waverly to visit with Glenn and Waverly Light and Power.*

we have a platform for doing something about global warming! I'm not talking about doing without energy. I'm saying let's provide cleaner energy and let's educate our customers to use energy smarter. A lot of the way business is set up in this country is counterproductive—that is, if the goal is to make money, then the solution must be to sell more product. In the utility business, the goal should be to sell our product and maximize that output in as environmentally responsible a fashion as possible."

During Glenn's tenure as manager, Waverly Light and Power was able to reduce its CO_2 emissions by 20 percent from a decade earlier. It's one of the reasons that during the 2008 presidential campaign in Iowa, then-presidential-candidate Barack Obama traveled to Waverly. While campaigning all over the state, he repeatedly heard about the small city that was making big contributions to energy efficiency. Addressing people in the Waverly Light and Power service garage in summer 2007, Barack Obama called for energy efficiency as a priority for the nation and held up Waverly Light and Power as an example of what was possible.

Glenn is clear about his reasons for supporting a voluntary reduction of greenhouse emissions. "We were able to demonstrate to our community and to our board of trustees that environmental commitment makes economic sense both now and in the future. As a community asset, our business must not focus myopically on short-term economic gain." Glenn is a businessman who planted trees that will shade Waverly for a hundred years and who is equally determined to plant policies that will sustain the community for a hundred years.

How CO_2 Reduction Works—Carbon Sequestration

As global climate change becomes a recognized phenomenon, "carbon dioxide reduction" and "carbon sequestration" have become buzzwords. The terms are often used interchangeably, though the former is the biological process and the latter is the desirable result. These complex terms are probably best understood by looking first at one tree—for example, the old burr oak in downtown Waverly. The burr oak is a deciduous tree, meaning it loses its leaves in the chilly, shortening days of an Iowa autumn. During spring, summer, and fall, the oak's leaves practice photosynthesis—that is, they convert carbon dioxide in the atmosphere into oxygen and various organic compounds with the aid of sunlight. The various carbon-containing compounds it produces, especially sugars, are essential for the tree's growth. All trees and plants are master "carbon dioxide reducers." They are performing this function during all daylight hours in all seasons that they have leaves or needles. Trees and other plants are a natural antidote to global warming. As part of the photosynthetic process, they "breathe in" carbon dioxide and "breathe out" oxygen for animals and humans to breathe. In simple terms, more trees mean less global warming.

Carbon sequestration is defined by Wikipedia as "a technique for the long term storage of carbon dioxide or other forms of carbon for the mitigation of global warming." Waverly's burr oak is a standing example of stored carbon. All of the trees and forests in that central Iowa town are sequestering carbon. However, if one of them dies or is burned for firewood, it releases its carbon into the atmosphere—thus adding to global warming. Planting trees sequesters carbon. Leaving a field of corn stalks standing during the winter sequesters carbon. Carbon sequestration also occurs when special scrubbers on coal and natural gas power plants remove carbon dioxide and store it in underground chambers. Burial of garbage in landfills is actually also a form of carbon sequestration.

The U.S. Department of Energy Web site is a good source of information on carbon sequestration. The site explains, "The global biosphere absorbs roughly two billion tons of carbon annually, an amount equal to roughly one third of all global carbon emissions from human activity." In plain English that means existing trees, other vegetation, and the ocean (another carbon sink) clean up one-third of all the CO_2 being produced by human activity. In further plain English that means that two-thirds of carbon production is not being dealt with naturally. There aren't enough trees, plants, or ocean to clean up the ever-increasing amount of CO_2 emissions from deforestation and burning fossil fuels. More trees and other means of carbon sequestration are needed, *and* emissions need to be reduced.

Glenn and his APPA counterparts in Tree Power are enhancing their carbon sequestration efforts by voluntarily doing everything from adding wind power (to replace CO_2-producing power production like coal and natural gas facilities) to planting trees to encouraging decreased energy usage. They are setting a course of change that is both visionary and

Photo Courtesy: Waverly Light and Power

Thanks to the efforts of Glenn and his board of directors Waverly Light and Power joined four much larger power companies to answer the challenge of the World Wildlife Fund to commit to increasing its energy efficiency fifteen percent by 2020. »

drastically necessary. Glenn, and fellow members of the APPA, are not waiting for regulation that requires them to make environmental change; they are leading utility companies into sustainable practices. APPA's Tree Benefits Calculator calculates how much electricity-generating capacity a utility can save by planting trees in specific locations and also estimates how many pounds of carbon and carbon dioxide will be sequestered by these same trees. These public utilities are voluntarily switching to cleaner technologies and utilizing the cleaning power of trees because it's good business. "As public utilities, we have a monopoly," explains Glenn. "Along with those rights come responsibilities. I believe we have a moral obligation to be good stewards of the resources we work with."

In 2004, Waverly Light and Power became one of five power companies to answer the challenge of the World Wildlife Fund to commit to clean energy and support limits on carbon dioxide emissions. The WWF PowerSwitch! Challenge requires utilities to support binding limits on national CO_2 emissions and undertake one of several action targets. Along with much larger companies like Austin Energy (TX), the Sacramento Municipal Utility District (CA), Burlington Electric (VT), and the FPL Group (FL), Waverly Light and Power committed to increasing its energy efficiency by 15 percent by 2020.

The Challenges of Trees

Trees provide utilities with a triple whammy for environmental stewardship, and they also challenge power companies. They grow, they age, they blow down. Major blackouts have been traced to the collision of a single tree with a power line or transformer. In August 2003, a massive blackout in the Midwest, Northeast, and eastern Canada left millions of people without power for days. After months of flinging blame back and forth across the border, and back and forth between the more than one hundred power plants that shut down, a February 2004 U.S.–Canadian task force said the main cause of the blackout came from failure to trim trees in Ohio.

Proper tree pruning is an important part of any utility company's maintenance program. Glenn doesn't see a problem with poorly placed trees—he sees an opportunity. Waverly Light and Power has a Tree Trade program. When the utility identifies trees that cannot be effectively pruned or that cannot coexist with power lines, the utility approaches the landowner with a proposal: If we can remove the tree, we'll plant a new one in a safer location. "In the long run, tree replacement saves us a lot of money," says Glenn. "For one of our large employers in town, the loss of even a few minutes of electricity requires so much restart recalibration in their equipment it can amount to

⌃ *Though Glenn and his wife, Ginger, are happily retired, Glenn still travels widely to speak with small municipal and rural electric cooperatives talking about the triple whammy investment of trees and the story of lowering customer bills and energy consumption at Waverly Light and Power.*

the equivalent of a month's worth of electricity to get it going again. And, of course, we don't even like to think about the dangers to nursing homes or our hospital."

In addition to the Tree Trade program, Waverly Light and Power has a tree-trimming program to keep lines clear. "We work hard to train our people on the correct way to trim trees. They do not simply hack the tops off trees growing into lines. One of our line foremen, Brad Schmidt, is a member of the local Trees Forever chapter and regularly attends the Trees and Utilities section of the National Tree Conference. He trains our crews. There is good craft to the proper trimming of trees."

Iowa and Beyond

When asked why he had remained general manager of this small utility in the middle of the country when he was offered higher profile jobs elsewhere, Glenn responds, "It was immensely gratifying to live and work for a community and a utility that are genuinely passionate about environmental stewardship. There are unique advantages about Iowa. People are generally very connected with the land because so many of them have come from and are still connected in some way to agricultural roots. They understand that taking care of the land and the environment is important because the land is and always has been their livelihood.

"When Iowa celebrated its sesquicentennial a few years back, Tom Brokaw reported that 25 percent of the most fertile farmland in the U.S. is located in Iowa. Waverly sits right in the middle of that tremendous resource. We are in a sea of corn and soybeans, and the Cedar River runs right through the center of town. The need for stewardship and the interdependence of field, trees, groundwater, and wind is all right there. Waverly Light and Power taught me big lessons on a manageable scale," says Glenn.

In 2007, Glenn retired, and he and his wife, Ginger, moved to North Carolina to be closer to their families. Still vibrant and knowledgeable, Glenn actively continues his tree stewardship and his pursuit of the importance of energy efficiency and alternative energy. In the work he is doing as a Clean Energy Ambassador for the Plains Justice nonprofit law firm, Glenn travels all around the United States working with small municipal and rural electric cooperatives, sharing the story of lowering customer bills and energy consumption at Waverly Light and Power. In a fall 2009 article in Spain's *Guide Post* magazine, Glenn is quoted extensively about the potential of wind energy in the future. Glenn made significant contributions to *The Energy Efficiency Guidebook* published by the Energy Center of Wisconsin as a comprehensive resource for community-owned electric and gas utilities. Working with the retired general manager from SMUD, Jan Schori, Glenn is also working with the Clean and Efficient Energy Program of APPA and the Alliance to Save Energy. "We're just a couple of retired tree huggers still out there doing our work," Glenn smiles.

When asked what advice he would have for people living in communities that are not so tree conscious, Glenn replies, "Well, I'd start by planting a tree. I've always remembered that old adage, 'The true meaning of life is planting a tree whose shade you'll never benefit from.' I'd also find out what environmental things are going on in your community. Most communities, especially large cities, have volunteer programs for helping with park maintenance. And wherever you are, get to know more about your utility company and what kinds of conservation programs they have. If they are not involved with planting trees or trading trees or carefully trimming trees, find out why not and start requiring them to become environmental stewards. Give them the triple whammy and show them the Tree Benefits Calculator. Public insistence is a great motivator."

Glenn's ability to stay inspired and enthusiastic while sitting behind a desk, delivering a PowerPoint Presentation, and managing figures is a long walk from the woods of his earlier years. Yet his father's and grandfather's tutelage show through in everything he has accomplished. Somewhere, it makes the old men proud.

Photo Courtesy: Marilyn Foster Hendee

Name: Dr. John Hendee

Occupation: Tree farmer and international wilderness activist, retired U.S. Forest Service, and university researcher

Point of Interest: Preservation of wilderness through research, education, and legislation is critical for the future of people in all countries.

CHAPTER 10

The Wilderness Researcher

Dr. John Hendee

I would encourage anyone feeling stressed by the pressures of twenty-first-century life to go sit under a tree, and all the better if it is in wild country or wilderness.

—DR. JOHN HENDEE

BENEATH AN ANCIENT GNARLED WHITE oak tree on a ridge overlooking the rolling, wooded hills of his North Carolina farm, a man sat silent and listening. Spindle trees of heavy green poked occasionally through the deciduous canopy. A summer breeze brushed his midlife face. Everything about the scene was peaceful—except the man himself. It was 1985. John C. Hendee was forty-six years old, a prolific wilderness researcher working within the U.S. Forest Service. He had already authored dozens of publications for the Forest Service on tree management, a number that would continue to rise throughout his professional life, but he was restless and uncertain how to proceed. For twenty-five years he had served the U.S. Forest Service—pioneering many of the scientific studies on managing public lands as wilderness. He was a second-generation Forest Service man and a third-generation tree farmer, and lately he had begun to feel an inner longing for change. As a hugely capable wilderness researcher, John might not have found these inner longings for change to be a welcome occurrence. But life, like wind in the oak's branches, shakes people loose until the acorn falls from the tree and finds its own place in the forest.

From the time he was a boy accompanying his U.S. Forest Service father into the high Sierra Mountains of California, John had found peace and joy in places where the sights and sounds of civilization faded away. His passage from boyhood to manhood centered partially on his increasing ability to walk farther, climb higher, and carry heavier loads. It was natural for him to travel to wild places to test his mettle and mature into the next phase of his life. On this day, John had gone in search of his old friend, the white oak.

John sat observing the tree for a long time. He noticed the girth and roughened bark of the tree. He observed the curving, arching, sweeping motion of branches. And he was calmed by the gentle rustling of wind through its leaves. In 1979, when his North Carolina farmlands were still new to him, he had attempted to walk every inch of what he owned. In the process he had discovered the old oak and slept under the branches of this sentinel tree. But in recent years, such hikes and reverie in nature had become little more than a distant memory while John, as assistant director of the Southeastern Forest Experiment Station in Ashville, found himself consumed with administrative concerns such as overseeing research deadlines and raising funds. Now, he was back. He was breathing deeply, calming down, letting go of busyness, and getting ready to listen.

"That time with the tree was really important," says John from the perspective of a quarter century later. "I felt like I relaxed into some new language for myself, and the tree spoke to me that what I was seeking in my work life was an opportunity to branch out." John hiked down from his day with the oak, carrying a resolve to actively seek new professional direction, to let himself branch out. First, though, he wanted to consult with his mentor and best friend—his father.

"I was struggling with the feeling that leaving the Forest Service meant turning my back on the family business," explains John. "My father, Clare Hendee, was a thirty-eight-year veteran and former deputy chief of the U.S. Forest Service. So, I needed to talk with Dad about the changes brewing inside myself."

The U.S. Forest Service was founded in 1905 as an agency under the Department of Agriculture. It manages 155 national forests and twenty national grasslands and oversees a total of 193 million acres. Gifford Pinchot, the first chief of the Forest Service, set the philosophical and practical applications for how the United States would steward these diverse and massive resources. In only five years, Pinchot reformed the management of forestlands toward a conservation ethic of planned use and renewal. He called this ethic "the art of producing from the forest whatever it can yield for the service of man." The term *conservation* continues to change people's minds and policies.

The donation Pinchot made was revolutionary for this time, and just as John, the son, was influenced by his father, Clare, Gifford, the son, was influenced by his father, James. James Pinchot had made a fortune in lumbering and land speculation and provided his family with wealth, estates, and endowment funds. But when he turned around and looked at what he had done to the wilderness, James was filled with regret. He determined his son, Gifford, would become a forester and bring a more conservation-minded ethic to the management of forested lands. James personally endowed the Yale School of Forestry in 1900 and turned the family estate, Grey Towers (now a National Historic Site), into a tree nursery for the forestry movement. Gifford's brother, Amos, took over the business aspects of their family money and freed Gifford to research and write, developing the land management concepts that the father and son Hendees would inherit. Gifford Pinchot led an active political and policy-setting life and died in 1946, the year before Clare Hendee started to work for "the Service."

As with many fraternal organizations, the U.S. Forest Service is carefully structured with a chain of command. The chief of the Forest Service reports to the under secretary for natural resources and environment within the U.S. Department of Agriculture, which is an appointee of the president confirmed by the Senate and seated on the Cabinet. There are five deputy chiefs, one of whom had been Clare W. Hendee, John's father. What happens in these large organizations at the top levels of administration and what happens at the level of the ranger in the field are very different but have in common an intense sense of bond around mission, service, management, and loyalty to the "brotherhood" (which now includes women and minorities in a way that was not historically present). It was the weight of this heritage that accompanied John down the mountain to talk with his father about stepping outside the world of organizational service he had been raised within and known for the first half of his working life.

⩘ *Clare and John Hendee carried on a long tradition of father/son teams working for the U.S. Forest Service.*

Early Years

John's passion for wilderness started at a very early age. "Because my dad moved around as a forester—Michigan, northern Minnesota, Oregon, Colorado, and California—I, as the only son, had the chance to go to the woods with him in many different places. When he was the regional forester in California in the early 1950s, he took me on several trips to the high Sierra Mountains where we would fish in a lake or stream.

"One of Dad's district rangers, Arnold Snyder, was managing a large area of the high Sierras as a total wilderness area even though it was well before the 1964 Wilderness Act. One memorable trip we rode several days with Arnold across his Sierra National Forest District into Sequoia National Park where Dad spoke at a Sierra Club gathering. The scenery, wildness, and fishing were great. That trip was my first introduction into really big wilderness, and I developed a fever for that kind of experience. I knew my destiny was to be a forester and follow my father into the U.S. Forest Service."

⚡ *John grew up exploring the high country of California's Sierra Nevada mountains. This love of wilderness guided all of his U.S. Forest Service work.*

Photo Courtesy: Ann Linnea

Like his father, John earned his bachelor's degree in forestry at Michigan State University. Although a generation apart, they had the same dendrology professor—Dr. Karl Dressel, author of the defining textbook on dendrology (the study of trees) at that time. In 1959, John missed much of a forest economics class because he was traveling as a player on the varsity baseball team. As part of his makeup work, he had to prepare a term paper. At his father's suggestion, he chose the topic of "wilderness as a forest policy issue."

The conservation mandate of the Forest Service at the time was tilted toward viewing trees as a long-term crop-rotation effort: harvest (i.e., cut down), plant, wait a few decades, harvest again. With a 193-million-acre field to plow, and a continent to oversee, the Forest Service felt this definition wasn't unreasonable. Wilderness was a by-product, and protecting wilderness was the responsibility of the National Park System—the Service was a sort of ranch for trees. *Multiuse* and *multipurpose* were the key words, and the stress of population demand was not yet at its peak—either for forest products or for forest recreation. Something called "the wilderness bill" had been introduced in Congress in 1956. It was in and out of committee and put up for vote, and it failed and was sent back to committee. The elder Hendee, who was closer to the department and political scene told his son, "This wilderness bill won't go away. It's going to be big. And eventually it's going to pass into law." Prophetic words—and a good term paper project.

Unknown to both of them, this one college paper would set a course that would guide John's entire professional career. It laid a foundation of knowledge on top of his already strongly developed love of wild places. With his father's help, John tackled the paper over spring break. He met with Reynolds Florence, the director of legislative affairs for the U.S. Forest Service, and reviewed all the news clippings about the hearings on the wilderness bill. And then he wrote the first of what would be many professional papers on wilderness, especially after the passage of the Wilderness Act.

The Forest Service

John graduated from Michigan State in 1960. His first job was reforestation on the Waldport Ranger District of the Siuslaw National Forest in Oregon. At that time the Waldport District had the largest allowable timber harvest of any district in the nation. "Nationally the wilderness issue was getting hot at this point," explains John. "But the he-man timber foresters on my district were pooh-poohing wilderness." These are civilized words for the struggle going on in the woods between competing interests over the trees. "Largest allowable timber harvest" means a lot of trees were being cut, and the boy who had hiked the high ridges was now at the lower altitudes and the lower echelon maturing himself into the rigors of life within the U.S. Forest Service.

During his years in Oregon, John completed his master's degree in forest management from nearby Oregon State University. He moved from reforestation to timber sale layout and appraisal and then logging inspection. Though he was learning the trade of timber harvest, John's real passion remained admiring the trees, not cutting them down. In his spare time he served as the local scoutmaster. "I remember taking Scouts on hikes in remote areas of the ranger district that were ultimately designated wilderness—that is, Drift Creek, Rock Creek, and other areas not yet penetrated by roads. My Forest Service boss would tease me, saying, 'If we had a timber sale there, you would have been able to drive instead of hike.' The prevailing attitude on the district was clearly the more area roaded and logged, the better."

As Congress labored toward passable legislation that would mandate an extended protection of wilderness and forests, it was clear that something was going to change when the bill became law. Many developers and competing interests—logging companies, land holders, and the Forest Service itself—pushed forward the kinds of infrastructure that would open land to potential harvest or multiuse: all it took was a gravel road. In 1961–62, John F. Kennedy was newly president, Orville Freeman was secretary of agriculture, the cold war was on, and the Cuban Missile Crisis was building—thus, public attention was focused on The Bomb, not the forest.

The Wilderness Act

And then, on September 3, 1964, President Lyndon B. Johnson signed the Wilderness Act "in order to assure that an increasing population, accompanied by expanding settlement and growing mechanization, does not occupy and modify all areas within the United States and its possessions, leaving no lands designated for preservation and protection in their natural condition." Basically, the law restricted grazing, mining, timber cutting, and using mechanized vehicles and equipment in formally designated wilderness areas managed by the U.S. Forest Service, the Bureau of Land Management, the U.S. Fish and Wildlife Service, and the National Park Service.

According to the law, wilderness is defined as "an area where the earth and its community of life are untrammeled by man, where man himself is a visitor who does not remain." The bill mandated that each of the four federal land management agencies map and define roadless areas within their jurisdiction that fit this definition and recommend which of them should be designated as wilderness. It was a revolutionary piece of legislation that set in place huge changes in these agencies.

The focus of the U.S. Forest Service has always been multiple-use management of national forests and grasslands. This means it manages these defined territories for timber, grazing, and recreation and supports research to guide this management. By contrast, the mission of the National Park Service, founded in 1916 and housed under the Department of the Interior, was to "preserve unimpaired the natural and cultural resources and values of the national park system for the enjoyment, education, and inspiration of this and future generations."

Before passage of the Wilderness Act, the National Park Serve had as its primary focus the job of protecting resources, as did the U.S. Fish and Wildlife Service. However, the Bureau of Land Management and the U.S. Forest Service had a twofold mission of both protection and use. When the Wilderness Act required the Bureau of Land Management and the U.S. Forest Service to set aside some of their lands as wilderness, where mechanized uses and logging would be restricted, this was contrary to their histories and to the mind-set of their managers and the chain of command. The new legislation created the need for readjustment and reapplication of the mission and created significant internal challenges. The new requirements needed a new voice. And though he would be too modest to say it, John Hendee was one of those voices. He was an advocate for wilderness with a research background that gave him credibility in the adjustments at hand. He was young, he was Clare Hendee's son, and he was in the right place at the right time to help his agency.

John left traditional timber management at the local ranger district level in 1967 to pursue his PhD in forestry with a minor in economics and sociology at the University of Washington. His dissertation focused on wilderness usage. He says of that time, "I stayed in Seattle for twelve years working for Forest Service research and did some of the first wilderness user studies." In 1976, having also won a national Conservation Achievement Award for his research, he was selected as a federal congressional fellow. He moved to Washington, D.C., and served on the staffs of Senator Frank Church of Idaho and Congressmen James Weaver of Oregon. After fifteen months, he began working on wilderness legislative issues in the Forest Service national office.

Asheville, North Carolina, was John's next assignment for the Service. He arrived in 1979 to serve as assistant director of the Southeastern Forest Experiment Station, overseeing Forest Service research units in four southeastern states. Out of the city, he could return to another tradition carried on from his father—owning a farm. "I wanted my kids to experience the hard work and connection to the land that a farm brings. We owned three small farms and raised some beef cattle, sorghum, and tobacco. It was hard work, and ultimately my wife, who had been

Photo Courtesy: John Hendee

⌃ *Taking the time to be in the wilderness with his own children, John found an antidote to years of working in cities on high level research.*

raised on a farm, didn't appreciate it. She told me once that she went to college to escape from being a poor dirt farmer and didn't expect to marry one! But the kids had the experience I had hoped for them. They had a pony, dog, cat, chickens, did chores and helped tend a big garden, and I felt it was good for all of us." They were busy years—and eventually they would lead him to the white oak and his day of reflection and reckoning.

Looking back, John recalls that his conversation about leaving the Forest Service went easily with his dad, who was by then retired from the Service himself. "My dad was, of course, just great. He acknowledged that the Service itself was changing and that I could do good things for it outside as well as in."

The Farms

John's interest in farms kept him outdoors even when his Forest Service job kept him primarily behind a desk. This interest is a multigenerational heritage as strong in him as the Forest Service. His maternal grandparents had homesteaded on a farm ten miles from Sault Ste. Marie, Michigan, near Lake Superior and Canada. His mother, Myrtle, the second-youngest child in a family of seven children, had been born and raised on this piece of land. Myrtle was no stranger to hard work. A true pioneer daughter, she helped with chores ranging from milking the cows to feeding the chickens to learning homemaking from her mother. She was educated in the one-room schoolhouse in the nearby town of Rosedale and went on to earn a teaching degree at the Michigan Normal School. In 1934, she married Clare Hendee, newly graduated from the School of Forestry. In the first twenty-five years of their marriage, they lived in seven different

⊻ *John and Clare Hendee and a 100-year tradition of a family farm that includes Christmas trees, berries, and other crops.*

Photo Courtesy: John Hendee

MICHIGAN CENTENNIAL FARM

OWNED BY THE SAME FAMILY
OVER ONE HUNDRED YEARS

MICHIGAN
HISTORICAL COMMISSION
SPONSORED BY CLOVERLAND ELECTRIC

places, but they returned to visit her family farm as often as possible: It was the hub of the family wheel. The children brought their children home to the farm.

John has fond farm memories of shooting targets with his uncle's .22 rifle and catching crawdads in the river with his sister. His grandparents would live on the farm until they passed away in 1946 and 1949 (John was ten years old). His uncles took over the farm until the last one passed away in 1958. When the farm was put up for sale, John's parents bought it as a place to regularly visit in retirement. They couldn't tend to animals and seasonal crops, but they wanted to utilize the land.

"What do retired foresters do?" asks John. "They start a tree farm." Total acreage of the farm at that point was 320 acres. Clare and Myrtle made regular trips to plant Christmas trees, tend a garden, and pick wild berries. As Clare and Myrtle returned for visits, so, too, did their children and grandchildren. John's children were the fourth generation to enjoy jumping in the hayloft, wading in the river, picking wild strawberries, and making jam and jelly with Grandma.

The American Tree Farm System encourages the sustaining of forests, watersheds, and healthy habitats through private landownership. It was only natural that the two stewards of public forests, Clare and John, would seek to do the same with their private forest. They brought their considerable skills to managing the farm as a place of growing renewable forest resources. And eventually the land received Michigan Centennial Farm designation by the Michigan Historical Commission for being owned by the same family for over one hundred years.

In 1985, at the age of forty-six, John left the Service and moved his family to Moscow, Idaho, where he became dean of the College of Forestry and Natural Resources.

He sold the North Carolina farms that kept him connected to the land during his years as a forest service researcher and shortly purchased the 560-acre Big Bear Ranch where he and his family restored the eroding wheat land by planting trees, building six ponds, and creating a ten-acre wetland. He managed this major land renovation while writing the second edition of *Wilderness Management: Stewardship and Protection of Resources and Values*, keeping up with the challenges of academic leadership, and raising his young family. In addition, he joined his father and colleagues Grant and Noni Sharpe in coauthoring the sixth edition of the textbook *Introduction to Forests and Renewable Resources*. "I've always had boundless energy and been a compulsive worker," says John. "It's been a strength but also a weakness when I worked myself out of balance with my other responsibilities." During this time of tremendous productivity, his marriage strained and eventually dissolved.

The coauthoring and publication of a third edition of *Wilderness Management* in 2002 and fourth edition in 2009 culminated John's contributions to wilderness management. Published by Fulcrum Publishing in Golden, Colorado, the book is still the only textbook on the subject. "It traces the history, philosophy, legal basis, and accepted practices and processes for designation and managing the U.S. wilderness system. It also looks at international wilderness issues."

After this tremendous output, John was still searching for ways to expand his experiential connection to trees and wilderness. There was a wildness waiting in him to be discovered.

Wilderness for Personal Growth

After nine years (1985–) in the dean's office, John became director of the Wilderness Research Center at the University of Idaho. This was his parachute—his chance to focus on wilderness again before he retired. For nearly a decade—beginning with his time under the white oak on his North Carolina farm—John had personally been reclaiming his own belief that experience in the natural world offers people the potential for inspiration, healing, and renewal. This interest in the use of wilderness for personal growth took John into the "softer" side of natural resource research. It was a time when new terms were coming into the field, terms like *ecopsychology*, which brought together the disciplines of ecology and psychology. As director of the University of Idaho's Wilderness Research Center, John landed a project designing a "wilderness

Photo Courtesy: Marilyn Foster Hendee

discovery" program for the federal Job Corps to see if a weeklong wilderness experience could inspire the Job Corps students and help reduce the early dropout rate. Ultimately it did. During the three-year pilot, he explored other programs and methods to help refine wilderness discovery. For that purpose

« *In wilderness questing John found an important source of spiritual inspiration.*

he attended an advanced wilderness leadership course in 1995 at the School of Lost Borders. His borders would never be the same.

The School of Lost Borders was founded by Steven Foster and Meredith Little for the purpose of taking people into the wilderness to spend time in solitude and fasting. It was one of the first opportunities made available to people from all walks of life to seek direction for their lives through the ancient practice of vision quests. Steven and Meredith taught their courses from the desert mountains bordering the Owens Valley in California. John came seeking professional training, but the experiential format of the school instead offered him opportunities not unlike the one he had with the old oak in North Carolina.

For example, John remembers spending much of one day on his first four-day solo watching a lizard catching flies and documenting the technique the lizard used to catch his meals. "At first, he didn't catch a thing. Then he moved over a ways and caught five flies. I realized it was exactly what I needed to do—reposition my perspective on the problem I was having at work. It's so simple yet so profound the wisdom that comes from extended time immersed in nature seeking deeper truths."

In addition to expanding his professional horizons, the school introduced John to Marilyn Riley. Marilyn, Steven Foster's sister, had been leading vision quests for nearly twenty years. She was a powerful influence in reminding John that when people spend time in the natural world, they often are able to see things that are inside them and articulate them in a way that's helpful. Marilyn modeled a spiritual connectedness with nature that had always been a part of John's life but was at least partly suppressed by the rigors and traditions of organizational life.

In 1997, John and Marilyn were married. And, of course, John made sure that one of her first trips was to the Hendee tree farm. Their first job at the farm was to work with shaping the young trees on their way to becoming Christmas focal points for hundreds of families.

"I've got precious memories from years of helping Dad with planting and shearing trees at the farm. However, shearing and shaping small trees for landscape use can be tedious and sometimes overwhelming," says John. "That first time Marilyn and I were shearing, I heard her talking a few rows over. I listened carefully and realized she was talking to the trees she was shearing.

"She would tell each tree how she was going to get it straightened out so it could grow up to be proud of how it looked, reassuring it as she lopped off an awkward branch or drooping limb," recalls John. "That was an eye opener for me. From then on shearing was a running dialogue between the trees and me." As had been true for John on his vision quest with the School of Lost Borders, the metaphor provided by time spent on the tree farm was helpful to his work. "Shaping up little trees gone wild" applied perfectly to his research taking at-risk youth into the wilderness.

International Work

John's wilderness work is the thread that has connected his life and interest from youth to retirement. Not surprisingly, as he gained experience, his influence broadened beyond the United States. "I encountered the World Wilderness Congress early in 1982 at a time when I was looking for creative ways to stay involved in wilderness." (This was about the time he was beginning to feel restless in his career with the U.S. Forest Service.) Here was an international, nongovernmental organization devoted to promoting wildland conservation and species protection. Every few years they organized a World Wilderness Congress, and their third congress was being held in Scotland in 1983. The Wilderness Congress is a public environmental forum and the only international event dedicated solely to wilderness conservation. "I contacted the congress secretariat, saying I wanted to come to the 1983 gathering. The secretariat, Vance Martin, then environmental director for Findhorn, an earth-centered community in Scotland, wrote back and invited me to come and present a paper."

John, who was still working for the U.S. Forest Service, applied for funding to attend. The request was denied, but John appealed the decision to agency executives in Washington. "You're sending many foresters to Europe for timber-related conferences," John says. "I'm the only one from the Service invited to this wilderness congress. Why are you denying me?"

John's request seemed to fall on deaf ears, so he notified Vance Martin that he'd be unable to attend. Vance told him not to give up. Soon word came to John from Washington that his funding to attend the congress was approved and would he please help write the U.S. secretary of agriculture's presentation to that Wilderness Congress. John was thrilled. This was the leap into an international wilderness affiliation that continues to this day for him.

At the 1983 congress in Scotland, John met Ian Player (brother of the famous South African golfer Gary Player), author and founder of the World Wilderness Congress. Ian asked him to organize the science program for the next congress. "Get lots of good scientists to attend. Don't ask for money or advice. Just do it," said Ian.

So, John organized a science program for the fourth World Wilderness Congress in Denver in 1987. Eighteen hundred delegates from seventy nations attended. Plenary sessions were held in Denver, and a week of scientific symposia was held at Estes

Park. Shortly thereafter, John was elected to the board of directors of the WILD Foundation, the organizers of the World Wilderness Congress and the only international organization dedicated entirely to wilderness protection around the world. He has been a director ever since, and he has helped organize the fifth, sixth, seventh, and eighth congresses in Norway, India, South Africa, and Alaska, respectively, and the ninth congress in Mexico in 2009. "Being involved with WILD and these congresses inspires me," John says. "I see that people of all nations grieve the loss of wild country, and many are eager to join efforts to save it."

Retired and Active

John retired as director of the University of Idaho's Wilderness Research Center in 2002 at the age of sixty-four. He remains active on the board of the WILD Foundation and oversees the *International Journal of Wilderness* as editor in chief. In his retirement he has also found time to co-author the fourth edition of *Wilderness Management: Stewardship and Protection of Resources and Values* and a seventh edition of *Introduction to Forests and Renewable Resources.*

He remains passionate about trees, forests, and people's connection to them. "The future of the world depends on forests and wild country. They provide renewable building materials and are crucial in combating global warming, preserving valuable gene pools, and providing ecological services like clean air and water. We need to pay attention to forests and wilderness for our own survival and for our own peace of mind.

Photo Courtesy: John Hendee

"I would encourage anyone feeling stressed by the pressures of twenty-first-century life to go sit under a tree, and all the better if it is in wild country or wilderness. Let the stress be drained away by the more normal pulse of nature. And if you really relax into it, nature will speak to you—if only by helping your own inner voice become clear."

Even in retirement both John and his wife, Marilyn, continue their lifelong work of wilderness advocacy. »

Photo Courtesy: Ann Linnea

Name: Dr. Robert Van Pelt

Occupation: Forest ecologist

Point of Wisdom: The new science of canopy ecology is unveiling the miraculously complex world of life in tall trees. There are still record tall trees hiding in remote forests all over the world.

CHAPTER 11

The Man Called "Mr. Tree"

Dr. Robert Van Pelt

There is a deep spiritual connection I feel in old-growth forests. They are simply totally amazing and inspirational.

—Dr. Robert Van Pelt

LOOKING MORE LIKE A RED-BEARDED gnome than a scientist, Dr. Robert Van Pelt can answer any question relating to trees—size, distribution, identification, growth patterns, contributions, physiology, and ecology—because he has literally devoted his life to studying and climbing them. As this book is going to press, he is studying and climbing redwood trees in the remote coastal highlands of northern California. Here the fog rolls in with welcome mist, obscuring the high canopy. These are trees that touch the sky, and as if to honor their dominance, the sky reaches down to touch them back. This will be his primary base for at least the next five years of research. The gnome is home.

Bob, as he prefers to be called, has been a research associate in forest ecology with the University of Washington since the late 1990s. In his professional life he seeks to understand what grows and lives high up in the tops of very large and tall trees. In his recreational life he travels all over the world to locate, climb, and measure the largest trees in existence. And in the rare occasion when you can find him at home, he is either analyzing the data he spends so much time collecting or enjoying any number of woodworking projects in various states of progress in his basement.

Bob's first book, *Champion Trees of Washington State* (University of Washington Press, 1996), took him to every corner of the state looking for the largest specimens of all native tree species and over seven hundred kinds of introduced trees. His second book, *Forest Giants of the Pacific Coast* (Global Forest Society and University of Washington Press, 2001), took him on much longer forays. "I drove over two hundred thousand miles and hiked in excess of seven hundred miles just to locate these 117 trees. And I would do it all over again in a second, if given the chance."

Bob is a renaissance tree man. His knowledge, passion, and skills about all things arboreal defy categorization. He's a respected scientist in the field of ancient or old-growth forest ecology, and his artistic skills enable him to draw these trees with an accuracy that surpasses that in photographs. And he is a mentor and colleague to Will Blozan (Chapter 6), who calls him "one of the most well-known, respected big tree hunters in the world."

Early Years

Born in Waukegan, Illinois, in 1959, Bob was one of six children. "Everyone except me still lives there. I go back and visit a lot, but I couldn't live there again. It's the West Coast with its mountains covered with thick forests that excites me." Bob graduated with a physics major and a geology minor from Northern Illinois University in 1981.

"From the time I was eight years old, I planned to follow in my brother Bruce's footsteps and get a degree in physics. He was my hero. But by the time I graduated, I had learned that continuing in physics meant I must really, really specialize, and by then I was becoming more interested in living systems."

So, Bob took a few years off from a formalized degree track and took classes like botany, environmental studies, and ecology. "I discovered my interest was in ecology because it combines all of the sciences." While taking those classes at the University of Wisconsin, he came across a copy of *Wisconsin's Champion Trees* by R. Bruce Allison. Searching out champions located within the city of Madison, Bob found that many of the trees had not been measured for years. He got involved in the Wisconsin champion tree program and eventually nominated a few new giant trees.

Still searching for his life direction after giving up his single-minded focus on physics, Bob headed west and worked as a cook at Sequoia National Park and then Olympic National Park. While cooking at Lake Crescent Lodge in the Olympic National Park in 1986, he learned that the national champion grand fir was located nearby. When he went looking for it, he discovered that it had actually been dead for several years. Calling the American Forestry Association, keepers of the *National Register of Big Trees*, Bob discovered the organization didn't know it was dead either. He also learned that Washington, with all its immense, old-growth conifers, was one of only two states that did not have a state coordinator to check trees and keep records current. Bob's life was about to take a permanent turn into the role of passionate tree man.

In 1987, Bob started the Washington big tree program and got a mailbox at the University of Washington. "At the time I got the job of state big tree coordinator, I was a nobody. I was a cook," Bob smiles. "But I wanted the job, unpaid though it was. I wasn't somebody working in a state forestry office that had just been handed one more thing he couldn't manage. I knew Washington State had extraordinary big trees, and I wanted to find them."

At that point he was still reluctant to enroll in graduate school because he had the idea that he would become a slave to a major professor and not be able to explore his own ideas. But the timing was right for this young man with his insatiable curiosity about old trees. He arrived in Washington State when most of Washington's immense old trees had neither been measured nor found, and he was considering the merits of advanced scientific training when Dr. Jerry F. Franklin, the dean of old-growth forest studies, had recently transferred to the University of Washington and was engaged in numerous research projects involving old-growth forest structure, including the installation of a 270-foot construction crane for the up-and-coming field of canopy ecology.

A New Scientific Discipline—Canopy Ecology

Dr. Jerry Franklin, professor of ecosystem analysis at the College of Forest Resources, University of Washington, pioneered studies in fire legacy and succession that showed the U.S. Forest Service that there was an important role for old-growth forests—that not all lands should be clear-cut. Dubbed the "guru of old growth," Dr. Franklin is another person with a lifetime dedication to trees. His stated objective as a politically active scientist is to "buy the best deal I can for forests and trees in a world dominated by humans."

The 1990s were an exciting time to be a graduate student working under Dr. Franklin. Bob was one of ten students attending weekly meetings and interacting with the latest information on the politics of forest management. Federal court judges had ordered land management agencies like the U.S. Forest Service to cease selling timber on lands designated by the U.S. Fish and Wildlife Service as critical habitat for the northern spotted owl. Logging came to a halt in old-growth forest areas in California, Oregon, and Washington, creating tremendous controversy in local logging communities. The U.S. Forest Service and the Bureau of Land Management

Photo Courtesy: Ann Linnea

⩘ *Bob with longtime mentor and friend, Dr. Jerry Franklin, "guru of old-growth trees."*

> ## Old-Growth Forests
>
> "Old-growth is a structural term that has to do with the physical makeup of the forest. Words that are nearly synonymous include ancient, virgin, primeval, over mature, climax, and pristine. Old-growth forests must have a wide variety of tree sizes (both in diameter and height), and must contain dead wood in the form of logs and snags."
>
> —Glossary definition in Van Pelt's *Forest Giants of the Pacific Coast*

were forced to embark on new courses of ecosystem management, so more research funds became available to study old-growth forests. Dr. Franklin and his graduate students created a hotbed of old-growth forest research.

Acre for acre, old-growth Pacific Coast rain forests contain more than twice the biomass of tropical rain forests. Since these temperate rain forests receive the bulk of their precipitation in the form of snowmelt and abundant winter rain, conifers like Sitka spruce, western red cedar, western hemlock, coast redwoods, and Douglas fir can photosynthesize and grow nearly year-round—allowing them to often achieve gargantuan proportions. And though these trees are impressive from the ground, it was the study of them at the canopy level that opened new frontiers for scientists seeking to understand their role in the ecosystem. The canopy is a veritable hotel for mosses, lichens, and even small-growing trees.

Canopy ecology has been slow to make its way into the realm of recognized science. In 1970, one of the first canopy research projects occurred in an old-growth Douglas fir forest in Oregon. Supported by the International Biological Program, ten tall trees were climbed and every branch mapped and estimated for needle mass and area. At that time, Bob was just eleven years old and still focused on studying physics; Dr. Franklin, however, was involved in the study. An enormous amount of data was collected, but it was before the era of fast computers, and the data were difficult to manage. Other researchers pursued similar goals in other forests and used techniques varying from canopy cranes to canopy walkways to climbing to collect the data. However, it was not until the 1990s that the emerging science of canopy ecology came into its own with increased funding, fast computers, and a growing number of enthusiastic scientists. In 1994, Dr. Franklin secured funding for a canopy crane in the southern Cascades at the Wind River Research Station.

A canopy crane is basically a large industrial crane erected above an old-growth forest. Researchers ride in the gondola over the forest making observations and

measurements above the trees, or they can be lowered down into the upper canopies. The presence of the crane, funds from the spotted owl controversy, and Dr. Franklin's renown brought in scientists to study various aspects of the Pacific Northwest's old-growth forests—the world's most structurally complex forests. In 1995, as a new PhD, Bob was in a perfect position to become a researcher on some of the crane projects. In the next decade he would author or co-author over twenty scientific papers on various facets of forest structure and canopy ecology.

Though the canopy crane is a good research tool, it allows study on only a very specific section of forest: the one where the crane is located. It was at this time that Bob met a young scientist named Steve Sillett, who had been studying lichens in old-growth forest canopies in Oregon. The two shared many interests, in terms of both forest structure and canopy ecology and also giant trees. Steve has since become a Humboldt State University professor and world expert on redwood canopy biology. With his penchant for seeking to know all he can about big trees everywhere, Bob reasoned he had to learn about becoming a tree climber. He works with others who helped pioneer some of the techniques used for getting ropes up into trees in ways that minimize damage to them. With his sturdy build, red hair, and beard, Bob looks like a character from Sherwood Forest as he aims his crossbow and sends a small line over the top of a strong, low branch. Next he hoists his climbing rope over the branch with the aid of the shot line. His every action attends to care for the tree, its vegetation, and himself. When Bob ascends the tree wearing his climbing harness and assorted gear, he disappears quickly into the permanent green canopy of his beloved temperate rain forest.

Combining Science and Big Tree Hunting

During the four years Bob worked on his PhD under Dr. Franklin, he completed his work on *Champion Trees of Washington State*. It was an ongoing balancing act. On the one hand, he was intent on the accuracy of the forest measurements he was taking for his PhD study of the understory tree response to canopy gaps in old-growth Douglas fir forests. On the other hand, Bob had became fast friends with Arthur Lee Jacobson, author of *Trees of Seattle*, and he spent much time measuring champion trees in the metropolitan area and then traveling all over the state in search of other large trees.

Part of how Bob managed the balancing act of PhD dissertation and the call of wild tree hunting was the support of his major professor. "Jerry Franklin has been my

« *On their way up to treetop level, students and researchers can get a bird's eye view of ancient, old-growth trees.*

professor, mentor, friend, and employer for more than a decade. He understood my obsessions with big trees and gave me some freedom and equipment to pursue them." So, contrary to the stereotype of being enslaved to a major professor's ideas, Bob found himself with a major professor who expanded and supported his dedication to trees.

"The best Washington community/capita I found with large, old, ornamental trees was Walla Walla," explains Bob. "The combination of the soils, the climate, and the settlement patterns in that southeastern Washington community has created a terrific situation for large trees. People there have an interest in the trees they have. The Whitman College campus and Pioneer Park [an Olmstead Brother's Design] contain some really significant trees."

Once he finished his book on champion trees of his new home state, it was only natural that he would be lured into searching for big trees all over the west coast of North America. To write *Forest Giants of the Pacific Coast*, Bob traveled up and down the U.S. and Canadian coast working with other scientists and amateurs with similar passions for finding big trees. The community of West Coast tree-measuring people included Wendell Flint, author of *To Find the Biggest Tree* and pioneer in estimating stem volumes for giant sequoias; Randy Stoltmann, big tree coordinator for British Columbia; Al Carder, author of *Forest Giants of the World: Past and Present*; Steve Sillett; and Arthur Lee Jacobson, the Seattle big tree man. Randy Stoltmann and Wendell Flint have both now passed away, and Al Carder will soon be celebrating his 100th birthday.

Bob and other members of the big tree community disagree with the American Forest registry of big trees on several issues. Under the American Forest system, points are given to a tree for girth, height, and crown diameter. "This is a useful system for many species, but it has drawbacks. . . . Moreover, I have always been interested in how tall trees can get and what the tallest of each one of the various conifers in the Pacific Northwest was. The AF's *National Register* did not help me with this. With few exceptions, a national champion is not the tallest known individual tree," wrote Bob in *Forest Giants of the Pacific Northwest*. So, when Bob worked with the Washington State Program of Big Trees and with his *Forest Giants* book, he listed more than one champion for each species. The book describes the American Forest champion tree of a given species and also describes the tallest, the stoutest, and in some cases the widest-crowned individuals.

Bob has taken the time to measure and study from three to ten of the most spectacular trees of every species represented in his book. He has listed ten Douglas firs for recognition. He has determined the volume, diameter breast height, overall height, and American Forest points for each tree. Each recognized tree is accompanied in this book by an interesting narrative. He wrote, "Only three of the ten largest Douglas-fir trees listed here had been discovered by early March of 1998—the Red Creek Tree, the Queets Fir, and the Indian Pass Tree. But the next six months saw more giant Douglas firs discovered than in the entire history of the Big Tree Program. The first of these spotted was Mauksa, which is easily visible from a major road. The Quinault area has so many giant Douglas firs that another 10-footer, about 250 feet tall, was no big deal. What looked to be a 10-foot-diameter tree, however, turned out to be nearly 12 feet, and the 250 feet was actually 290 feet. We stumbled upon Mauksa while searching for the still larger Tichipawa. This tree is so close to the road that it still amazes me that no one had noticed it before—and so much so that it seems to laugh at me whenever I drive by. We named it Mauksa, which is Quinault for 'one who laughs all the time.'

"Like my friends in the East—the ENTS [see Chapter 6]—I am more interested in what a tree can achieve than how it fits into some point system," explains Bob. "Measuring trees is a really, really meticulous process. There are only a handful of people in the world whose measurements I would trust. Will Blozan's numbers are impeccable."

The Scientist Artist

Walking through an old-growth forest with Bob, the man whose car license plate reads "Big Trez," is an invitation into greater acquaintance with and reverie for trees. Listening to him speak, one begins to understand the artistic side of this man. "There is a deep spiritual connection I feel in old-growth forests. They are simply totally amazing and inspirational. For example, redwoods are as close to immortal as trees come. When a windstorm sheers off the tops, they just grow new ones. They remain the tallest living things on earth, growing to over 370 feet in height. Though sometimes very little grows on the forest floor beneath them, their canopies are incredible. A tree in the Federation Grove often has a pool of water in a trunk cavity where the top had broken off in the past. Sword ferns, huckleberries, and other smaller epiphytes are common in these giants. Another tree has about two meters of soil in the main crotch, a spot where the top had sheared off centuries ago and has grown back."

The scientist and the artist are all balanced here. Bob's insatiable desire to understand and communicate everything honed an early ability to draw. The artistic renditions of trees in Bob's two books are so exquisite people buy individual drawings as artwork.

Illustration Courtesy: Robert Van Pelt

Several skills enable this scientist artist to make "architecturally correct" drawings. First, he takes on-site, laser measurements to ensure proper proportions. Second, he makes sketches in the field. Finally, he photographs the big tree he is studying. In dense forests he must take photographs from many angles to capture the specifics of each tree. Once home, he combines these three pieces of information to create stunning portrayals that honor the remarkableness of old-growth trees. His drawings are used widely in scientific texts and talks.

A fine example of Bob's drawings is that of the General Sherman Tree, the single largest living tree in the world. This sequoia (sequoias are different from coast redwoods in that they live in the pine and fir forests of the Sierra Nevada) has no peer. In Bob's words, "Since the giant sequoias were discovered about the time of the Civil War, early big trees were named after heroes from that war. General Sherman has been challenged numerous times, but with a diameter of 27.1 feet and a volume of 55,040 cubic feet, it is the earth's largest living thing—weighing nearly ten times as much as the heaviest whale."

His reflections on the most important timber tree in the world—Douglas firs—are those of both a scientist and a naturalist. "Dougs are remarkable.

⌃ *Drawing of the General Sherman Tree—a giant Sequoia, the largest living trees in the world.*

They lose lots of branches in the wind but are capable of epicormic branching, so they easily make new branches out of the wound in the cambium. Douglas fir old-growth is one of my favorite ecosystems because there is so much happening both on the ground and up in the canopy. And the studies being done on the Wind River crane are showing us that old-growth forests are largely carbon sinks—that is, they absorb carbon—unlike young plantations which over many cycles end up releasing carbon into the atmosphere. [Carbon sequestration helps mitigate the effects of global warming—see Chapter 9.] The remaining as yet undiscovered big Doug firs to be found are probably in the Queets rain forest on the Olympic Peninsula."

The Queets Rain Forest

Before his call to the redwoods project, Bob was involved in a three-year study of old-growth forests near the Queets River in the Olympic National Park. The study was funded by the Fisheries Department at the University of Washington. "It's very strange that fisheries people would pay me to climb trees," smiles Bob. "Needing to

Photo Courtesy: Ann Linnea

≈ *Whether climbing or taking a crane, canopy ecologists get close-up views and measurements of treetops that have radically changed what is known about trees.*

know how biomass was partitioned in the forest is the short answer as to why they've funded the program."

Biomass literally means "the weight of all life." Plant biomass refers to the weight of plant life in a forest—it is a measurement of the amount of photosynthesis generated by that particular forest—minus the respiration of the living cells. Fishery biologists are interested in understanding the complete nutrient cycle in and around salmon rivers. Since trees and plants contribute nutrients to the rivers through windfall and decay, if scientists can know the plant biomass surrounding those rivers, then they can better understand the nutrient cycle and the river's health from a salmon's perspective.

Bob and his assistants spent a year mapping a specific section of the forest surrounding the Queets River. Then they climbed key trees, collected and dried epiphytes (plants that grow on other plants), and measured the volumes of those key trees. "I think they hired us because we can measure tree structure, identify existing epiphytes, and estimate epiphyte mass faster than anyone else," he explains. "Furthermore, through hierarchical subsampling we can extrapolate beyond the tree level to the whole canopy."

Once Bob and his crew collected the epiphytes, they took them to the laboratory to be identified, dried, and weighed. Common epiphytes include ferns, lichens, and mosses. To give a perspective on the challenge of identification, nearly fifty-six species of mosses and liverworts along with several dozen species of lichens grow in the crowns of old-growth trees along the Queets River. "There are often over one hundred large branches in each of these trees all covered with mosses and lichens which when combined can contain up to a metric ton of dried biomass. Quantification of the amount of biomass will give fisheries people the data they need, and it is a dream come true for me to better study the epiphytes in the big trees."

Once they finished all the oven work (drying the epiphytes collected from the treetops), they discovered that the standing biomass of these huge, old, moss- and lichen-draped trees is higher than that of any other forest on the planet.

A research project the magnitude of the Queets study often requires camping and working under arduous conditions—especially in the rainy season. There are no nearby communities with restaurants or motels. When Bob is in the forest, he is totally immersed in his work, simply another creature living there. On one study trip he and an assistant accidentally hit a grouse as they were driving the back roads to one of their study plots. Emerging from the "Big Trez" Subaru station wagon, they looked sadly at the creature they had inadvertently killed. Following the ethics of

⋩ *Bob's vehicle traverses tens of thousands of miles looking for big trees in remote areas.*

the forest, they decided to let no creature's life be wasted. They picked up the grouse and later cooked it for supper at their campsite. Even under camping conditions of rain so steady that the Coleman stove required special coaxing to light, Bob the chef conjured up a gourmet meal.

Future and Current Projects

A big tree hunter is always looking for the next big quarry—which fortunately is not moving. And sooner or later a western big tree hunter is destined to end up climbing a redwood or a sequoia. In August 2006, on invitation from two fellow big tree hunters who have been systematically searching old-growth redwood forests for tall trees, Chris Atkins and Michael Taylor, Bob headed to Redwoods National Park for a tree-measuring expedition. They measured two new world record trees over 370 feet tall on a single day. While still in awe, Bob reported out to colleagues, "Without a doubt, this was the most significant day in history when it comes to tall tree measuring—and I am super-stoked that I was there to see history made! I am still completely baffled at how these trees could be so tall while on a steep slope so high above the creek, when all of the other tall trees are on flats! Just amazing! WOW!!!"

One of these trees was measured to 375.9 feet (later climbed and verified), and another dead topped tree measured to 371.2 feet. These trees surpassed the eight-year reign of a 370-foot giant discovered by Chris Atkins just ninety miles down the road at a state park. Better yet, a month later, they beat even those records with a nearby isolated tree that measured to 379.1 feet!

Beginning in fall 2009, Bob became a full-time scientist on a project to study the effects of climate change on redwoods and giant sequoias throughout their range. Save the Redwoods League, an organization devoted to redwood and sequoia preservation since 1918, is funding this project as part of its initiative to "manifest the most comprehensive and integrated redwood discovery period in U.S. history." One of the more spectacular projects launched as part of this initiative was to partner with the National Geographic Society and the Wildlife Conservation Society to sponsor the Redwoods Transect. From fall 2007 to fall 2008, National Geographic explorer Michael Fay hiked the 1,800-mile range of the coastal redwoods, bringing to public awareness both the magnificence of these remarkable trees and the possibilities for their future management. The Redwoods Transect is reported on at length in the October 2009 issue of *National Geographic*, and Bob provided the pen and ink sketches for this article.

Bob's work with the redwoods will focus on the creation of many permanent plots for future monitoring. Within the plot area, all trees will be mapped and measured, and on a subset of these, physiology measurements will be taken, dendrochronology cores studied, and transect plots of surrounding plants documented. "One of the first plots we established in Redwoods National Park is really brutal to get into," Bob admits. "It's a 2.5-kilometer bushwhack into the tallest upland forest in the world."

In addition to doing his research as a scientist and chasing tall trees, Bob is already working on his next book on forest giants of the world. "My book will document the individual trees that are the largest in the world," Bob explains. "It will be an accurate look at all the living large trees."

To find and measure all of the world's large trees, Bob does an extensive amount of traveling. On the modest income of a research associate, he manages some of his travel by participating in or leading trips for the International Dendrology Society (IDS). Members of IDS are committed to the study and preservation of trees and are largely European and Australian based. On a recent IDS trip, he had an opportunity to see the world's largest baobab tree in South Africa. "We needed a 120-foot tape to wrap around its trunk. They are very unusual trees. You can completely girdle the tree, and it won't die. This and the eucalyptus are the two largest hardwoods that will make it into my world giants book."

In Australia Bob's search for giants has focused on the fastest-growing trees in the world—the eucalyptus. Australia has over seven hundred different species of eucalyptus trees. In the particular forest Bob is studying, eucalyptus can grow taller than three hundred feet in one century.

"They are huge trees with no epiphytes in the canopy," explains Bob. "And there is no middle canopy. It is a structurally simple canopy unlike the forests of the Pacific Northwest. They have the unique situation of these large marsupials [called gliders] that eat the leaves. The largest things in our temperate rain forest that eat foliage are things like aphids and caterpillars. We have done some interesting work climbing these trees and improving our protocols for tree mapping."

In February 2009, Bob traveled to Tasmania in response to the LiDar (light detection and ranging) report from a friend that indicated the location of a eucalyptus over three hundred feet tall. LiDar is a remote sensing machine that can bounce signals off the ground and canopy from an airplane that then produces a 3-D view of whatever is below. Bob and colleague Steve Sillett climbed the tree and confirmed it to be 329 feet tall—the new tallest tree in the southern hemisphere.

Other countries Bob will visit while collecting data for his world giants book include Chile, Taiwan, Turkey, and Japan. Though already one of the recognized world experts in finding and measuring large trees, Bob continues to believe the work of learning about "the most magnificent living things on the planet" is still beginning. And for him, the goal of sharing what he learns about the world's old-growth forests through writing, teaching, studying, photographing, and drawing has also just begun. It's a whole different life than physics!

Photo Courtesy: Ann Linnea

A generous teacher as well as a researcher, Bob seldom passes up an opportunity to share his knowledge about trees. »

Photo Courtesy: PlantAmnesty

Name: Cass Turnbull

Occupation: Gardener, writer, activist

Point of Wisdom: Tree planting has become both trendy and politically correct, but the fate of these trees is often less promising. An incredible number of trees and shrubs are wasted because of ignorance about watering, mulching, pruning, and siting.

CHAPTER 12

The Activist Pruner of Emerald City

Cass Turnbull

When I started PlantAmnesty twenty years ago,

it was grim for big trees in Seattle.

It was only a matter of time before they were topped,

but we've largely managed to stop that.

—CASS TURNBULL

CASS TURNBULL IS A FLAMBOYANT pessimist, part stand-up comic, and a full-out advocate for the good life of city trees. She's a sturdy, not-yet-gray, midlife woman who can wield the tools of her landscaping trade and trade repartee on the local NPR radio station. Botanically opinionated, she makes people, laugh, cry, and join her in taking responsibility for the urban forest of her hometown, Seattle,

Washington. Her opinions are backed by careful botanical facts, research, and experience. She is enthusiastic about emphasizing the negative. "Talking about the need for proper tree care, the need to prune right, the need to water—all those positive things—doesn't get through to people. They have no concept of the sheer immensity of the problem," she states. "Trees just stand there and take it—whatever we do to them. So, I started PlantAmnesty twenty years ago as a way to grab people's attention and to end the senseless torture and mutilation of trees and shrubs."

By PlantAmnesty's definition, tortured trees and shrubs are those that are improperly placed, watered, or staked. Mutilated trees and shrubs are those that have been improperly pruned into mere shadows of their potential. For Cass and other members of PlantAmnesty, seeing tree toppers at work or watching heavy equipment carelessly moving around the base of trees shows both ignorance and complete disrespect for the value of trees and shrubs.

Listening to Cass, it's hard not to imagine her wearing a cape, jumping out of a racy green Treemobile, pruning shears in hand, and leaping between a defenseless tree and some doltish homeowner or tree-service lackey. In reality, she drives a silver 1990 Ford Ranger pickup truck with handmade sideboards, toolbox, and tool holders for brooms, pruners, rakes, and shovels in her day job as sole proprietor of a landscaping and maintenance business. Often accompanied by a horticulture student to help with the heavier work, and gain quite an experiential education, she handles the rigors of self-employment and then volunteers about thirty hours a month for her nonprofit. In Seattle, plants can grow year-round with little attention to seasonal boundaries, and in Cass's life, her devotion to plants shows little boundary between her work and her volunteerism. What does a pruner do for fun? In Cass's case, she prunes. She talks about pruning. She educates, cajoles, shames, and badgers people into proper care of their trees.

"PlantAmnesty is a maintenance organization," Cass explains. "All our activities focus around the simple notion of treating trees and shrubs properly. That's not very sexy or compelling given the magnitude of the abuse problem, so we jazz things up with zany parades, protests, games, and terrific education programs." As president of the organization, Cass herself is responsible for many of the education programs for garden clubs, school groups, and weekend adult classes. To listen to her teach an introductory or advanced class on pruning is to be guaranteed some good laughter. There is always an outrageous story about some pruning horror, often accompanied by slides. Right now the PlantAmnesty Web site includes a pop quiz on basic tree knowledge that can humble most Saturday gardeners and a request for someone offering their yard for a PlantAmnesty "extreme makeover."

Says Cass, "The typical pruning horror is beheading, but bondage is a close second. And one of my personal pet peeves are quasi-topiary pruning shaped as buns or boxes or whatever cute shape shrubs can be forced into."

Early History

Cass is herself a rare species in the city—a Seattle native. When Cass was a Seattle third grader, her family moved from a neighborhood with a lot of kids to a neighborhood with few kids. "I was suddenly very lonely. We had a beautiful yard surrounded by a couple of empty lots with parks nearby. I bonded with nature to assuage my loneliness and keep myself amused." Cass spent time climbing trees in their orchard, enjoying the treehouse her father built, chasing the cats in the tall grass, catching snakes, and exploring empty lots.

Cass also had an early mentor who loved trees. She recalls visiting her Uncle Norman and his horses. The man saw in young Cass a receptive audience for his fury over all the clear-cutting being done in the Cascade and Olympic mountains in the 1950s and '60s. "Uncle Norman's stories became my repository for all the sadness of my childhood," Cass recalls. "I just cried and cried and for years afterward fixated on the destruction of the forests of the world."

As a liberal arts major at Fairhaven College—in Western Washington State University in Bellingham, Washington—she turned her back on this burden and spent those years as a party girl. "I could have cared less about nature, and I was a total physical wimp. I could be beaten in arm wrestling by a twelve-year-old." When Cass left Fairhaven, she realized she "wasn't qualified to do anything." She inquired about job possibilities under CETA (Comprehensive Employment and Training Act, begun December 1973), a jobs program that included matching women to nontraditional jobs. She landed a job with the Seattle Parks Department and discovered she liked physical activity. Her early love of the outdoors kicked back in, and she tackled jobs like mowing, chainsawing, and weed whacking with gusto. Then she met a Parks Department horticulturist, Andrea Furlong, who taught her about plants.

"Andrea and I dug out a garden on Queen Anne called Parsons' Garden. Unlike many of our city parks, this one did not have ball fields. It was a passive-use park with special plants and bulbs growing between stepping stones and an old wisteria vine." It was this secret garden that reawakened Cass's deepest passion for gardening and helped her realize that after eleven years of working with the Parks Department, she was ready to start her own business.

Stepping into Activism

Before leaving the Seattle Parks Department in 1986, Cass took every class she could. She is now a Washington State certified landscaper, a certified arborist, a veteran King County master gardener, and a graduate of many horticulture classes at the University of Washington's Center for Urban Horticulture. The more she learned, the more irritated she became with the plight of the city's trees at the hands of bad pruners.

"After Andrea pointed bad pruning out to me, like topped trees and unruly water sprouts on shrubs, I began to see it everywhere. People actually seem to go out of their way to kill trees. They think in the short term—how can I improve my view? They spend money to get their tree topped, which seriously shortens its life and makes it ugly."

This irritation with the shortsightedness of tree topping and other bad pruning festered inside Cass. She had the opportunity to take a self-improvement seminar through a program called the Forum. "It was one of those self-actualization deals— enlightenment in a bag," Cass quips. Her first homework assignment was to list all her complaints. Then she was to choose one and design a solution. "I randomly picked,

Photo Courtesy: PlantAmnesty

⩘ *Cass and PlantAmnesty have been teaching proper pruning techniques in Seattle for over two decades.*

'I hate bad pruning.'" The solution she brainstormed laid the framework in place for what would become PlantAmnesty. Her next class assignment was to take some action. She put together a slideshow of pruning horrors to share with classmates. It was an instant hit, and PlantAmnesty was born.

Informing the Public

Urban trees are especially susceptible to poor arboreal practices like root compaction, improper siting, minimal watering, and topping. The American Forestry Association states that a tree with an average life span of eighty years in the wild is expected to live about twenty years in the suburbs and only seven years in urban settings. In PlantAmnesty's pamphlet *Six Ways to Kill Your Tree*, urban tree owners are taught to avoid the most common pitfalls of poor tree care. The tone of the writing, like Cass's oral presentations, is directive, educational, and delightfully graphic:

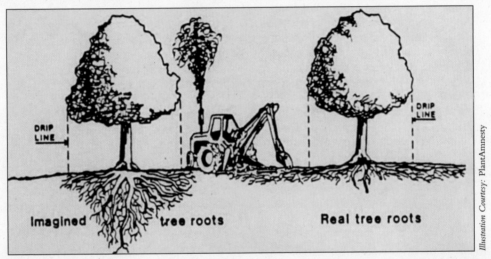

Illustration Courtesy: PlantAmnesty

1. *Do not forget to water, especially during the first two years and during droughts.* Water deeply with a long, slow trickle from the hose.
2. *Do not trench, cover up, or compact the soil in the root zone.* A tree's roots are shallower and broader than generally believed. They are not built like carrots. Tree roots need air and water and empty spaces in the soil.
3. *Do not leave tree stake ties, which can girdle the tree.* With staking, less is better. Remove ties as soon as the tree can stand on its own—one year, for most trees. Leave stakes to protect from mowers and bumpers for a while longer.

Illustration Courtesy: PlantAmnesty

4. *Do not plant a big tree in a small space.* Find out how tall and wide your tree species gets and give it that much room. Never plant tall trees under wires. This dooms them to mutilation and certain death.

5. *Do not top your tree or make repeated heading cuts (cut branch tips).* Topping and pruning to keep trees small causes them to rot and starve. Topping removes a tree's nourishment source—the leaves that manufacture its food. Besides killing the tree, topping it or cutting branch tips doesn't even work to keep it small. Ironically, it has the opposite effect: it causes rapid and unruly regrowth, which is not only ugly but significantly weaker than the original limbs.

Illustration Courtesy: PlantAmnesty

6. *Do not string trim the bark or bash the trunk with the mower.* The most living and vulnerable part of the tree is just under the bark. Trees die in slow motion from a series of blows over time.

Trees injured during construction generally give out five to ten years after the injury. With a little knowledge, we can create a kinder, gentler world for our friends the trees.

All of the protocols from the PlantAmnesty brochures come from the International Society of Arboriculture, the Tree Care Industry Association, and numerous professional publications, including those of Dr. Alex L. Shigo, of Shigo and Trees, Associates in Durham, New Hampshire. Dr. Shigo, a world-renowned scientist and author on the subject of arboriculture (trees), has dissected more than fifteen thousand trees and published over 270 articles to show why topping (also called heading, stubbing, or dehorning) is bad for trees. He has proved that the common practice of flush-cutting limbs off trees is also bad and that painting tree wounds with paints or sealants does no good.

One of the main educational goals of PlantAmnesty has been to get people to stop topping trees. In *Five Reasons to Stop Topping Trees*, the organization lists topping as the greatest threat to urban forests, despite the fact that it is a long-standing practice in many areas.

Five Reasons to Stop Topping Trees

1. *It won't work.* Topping won't work to keep trees small. After a deciduous tree is topped, its growth rate increases. It grows back rapidly in an attempt to replace its missing leaf area. It needs all of its leaves so that it can manufacture food for the trunk and roots. It won't slow down until it reaches about the same size it was before it was topped. It takes at maximum a few years before your tree returns to near its original size.

2. *It's expensive.* A topped tree must be done and redone every few years and eventually must be removed when it dies or the owner gives up. Each time a branch is cut, numerous long, skinny young shoots (suckers or water sprouts) grow rapidly back to replace it. They must be cut and recut, but they always regrow the next year, making the job exponentially more difficult. Topping also reduces the appraised value of your tree. A tree, like any landscape amenity, adds to the value of your property.

Illustration Courtesy: PlantAmnesty

3. *It's ugly.* The sight of a topped tree is offensive to many people. The freshly sawed-off limbs are reminiscent of arm or leg amputations. And that's just the beginning of the eyesore; the worst is yet to come, as the tree regrows a witch's broom of ugly straight suckers and sprouts. The natural beauty of the tree's crown is a function of the uninterrupted taper from the trunk to ever-finer and more delicate branches and the regular division of the branches. Arborists consider the topping of some trees a criminal act, since a tree's ninety-year achievement of natural beauty can be destroyed in a couple of hours.

4. *It's dangerous.* According to Dr. Shigo, topping is the most serious injury you can inflict on your tree. Severe topping and repeat topping can set up internal columns of rotten wood, the ill effect of which may show up years later in conjunction with a drought or other stress. Topping creates a hazardous tree in four ways:

    It rots.

It starves.

It creates weak limbs.

It creates increased wind resistance.

5. *It makes you look bad.* Topping makes you appear to be a cruel or foolish person. Your friends know better. You may top a tree to create a water view, but you should know that friends and neighbors see a view of a butchered tree with water in the background.

Cass and her cohorts at PlantAmnesty are happy to use social coercion and shame if necessary to save trees. And they admit that the source of the problem is twofold: In a city like Seattle, a "view" is considered water and mountains, not trees. "We have to create a little more identification with trees," she says. "That's why we talk in very biological language—beheading, cutting off arms, mutilation. That big green thing in your yard, hey, it's somebody with a life you can either support or ruin."

Basic Tree Biology

In addition to some of PlantAmnesty's outrageous tactics, one of the keys to the group's success has been its focus on scientific accuracy. Cass believes that helping people understand the inner workings of trees enables them to become better caretakers of them. PlantAmnesty brings home the message that everything about trees takes time—especially exhibiting external signs of damage.

A tree trunk has four primary parts: bark, the cambium, the sapwood, and the heartwood. The bark on each tree is unique and one of the key ways to identify conifers. In winter when the leaves are gone, bark is used to identify deciduous trees.

The bark on trees is like the skin on an animal: it forms a boundary between the inner and the outer. Bark provides extensive protection from fire, insects, and mechanical injury. The cambium, just under the protective bark, is where all the action in a tree takes place. The cambium is a single layer of living cells producing xylem on the inside and phloem on the outside. The xylem, which is made of thick-walled woody cells, is produced on the inside edge of the cambium. Its function is to transport water up and down the tree trunk from roots to leaf tips and to provide support. The thin-walled phlo-

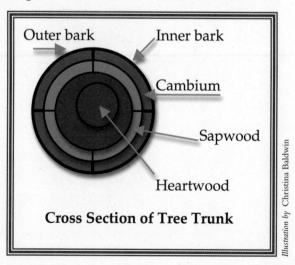

Cross Section of Tree Trunk

Outer bark

Inner bark

Cambium

Sapwood

Heartwood

Illustration by Christina Baldwin

em is located on the outside edge of the cambium, just under the bark. It provides a pipeline for liquid food for the entire tree.

Over time the xylem in the trunk becomes stiff with lignin to form the sapwood. As the cambium moves outward with increasing age, the living sapwood gradually clogs up with gum and pitch and becomes heartwood. *The most living, most vulnerable part of a tree is the green cambium just under the bark.* People wrongly assume that if they have nicked a tree trunk by backing into it, driving nails into it, or weed whacking it, they have done little or no damage because the tree keeps on living. The damaged tree may outwardly look fine for several years, but everything about trees takes time.

What is actually happening after an injury to the trunk is that the tree seals off the injury and simply keeps growing around it. Therefore, if the injury completely encircles the tree—for example, weed whacking—the phloem has been destroyed, and the tree cannot send nutrients up to the part of itself above the ring of injury. The xylem with its woody lignin is stronger and more resilient to injury, but the phloem is fragile. Over time a girdled tree, unable to send nutrients from the roots to the leaves and back to the roots, will die.

Similarly, if a limb rots or is snapped off, the tree surrounds the exposed wood with a wall that stops the microbes from invading the rest of the tree. What happens in the case of a topped tree is that the entire xylem and phloem system and the sapwood and heartwood of the tree are open and vulnerable to microbes entering the system.

Generally, though, trees die from a series of blows over a long time. As Cass wrote in an article for the *American Horticulturist*, "You may have a giant column of dead rotten wood walled off inside your tree from some old wound and when the drought hits, whammo! You've got a dead tree."

Proper Pruning and Thinning

On the basis of a scientific under-standing of trees, PlantAmnesty has created a strategy to educate prop-erty owners and professional tree-care people about the proper way to prune branches and trees. First, PlantAmnesty

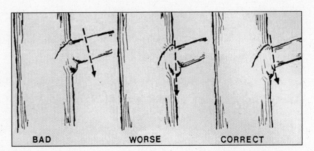

Illustration Courtesy: PlantAmnesty

decided to get people's attention by producing the pamphlets that emphasize the DON'TS. Then it prepared pamphlets that emphasize the proper SHOULDS.

For years, "flush cutting" was thought to be the proper way to prune tree branches. However, the work of Dr. Shigo and his fifteen thousand tree biopsies has shown that each branch has a bulge or support collar below it that is actually composed of trunk wood. "It's like a tree muscle," says Cass. "You can feel this bulge with your hand and especially on a larger limb see it like an armpit right under where the branch lifts off from the trunk. Now we know more than ever how important it is to carefully cut only the branch wood and to avoid cutting the collar because that is actually part of the trunk. Cutting the collar opens the trunk to rot and is responsible for serious problems that may show up later."

Maybe it's her humor, maybe it's her sheer determination, maybe it's that she likes to gesture at audiences with sharp implements in her hands, but Cass does seem to be getting through to the citizens of Seattle. "Mountains are great, water is lovely. What's wrong with looking through the lovely filtered branches of a tree or two? That's what I'm trying to get across to people; there are ways to have it all in nature—if we know how to care for what's around us."

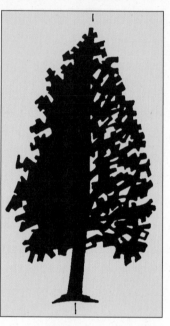

Illustration Courtesy: PlantAmnesty

Homeowners concerned about trees blocking their vista have four options: thinning, windowing, skirting, and selectively removing. Thinning trees (illustration at right) keeps their structure intact and takes out some of the view-obstructing foliage. The actual amount of foliage that can be removed varies with the species. As is shown in the figure, removing less than one-quarter of the total leaf area opens the tree visually.

The process of windowing works best when a large, close-in tree blocks the longer view. In windowing, a skilled arborist removes key branches to frame the view of what lies beyond. When you window a tree on one side, you may want to balance it with some pruning on the opposite side.

Skirting involves removing lower limbs. Depending on your perspective, you may be able to open up a view by removing the lower limbs of a tree. Also called crown raising, skirting is never done more than one-third the tree's visible height.

From a PlantAmnesty perspective, all these pruning techniques incorporate trees into the view and the aesthetic of the property, what grows close in, and what is enjoyed at a distance. However, if homeowners cannot abide having trees interrupt key parts of their view, PlantAmnesty always recommends complete tree removal rather than topping. "It's more honest," says Cass. "Don't try killing it softly."

And then there are power lines. According to Cass, "There are no nice solutions to the problem of trees growing into power lines. Utility workers get yelled at if they let trees grow into wires, they get yelled at for topping, they get yelled at for directional pruning and for taking the trees out. There is no reasonable way to keep a big tree small." PlantAmnesty supports careful utility workers who must deal with the fact that most wires are uninsulated, that thousands of miles of wires have big trees planted under or near them, and that every solution creates another problem.

A Good Cause Releases Energy

The formation of PlantAmnesty in October 1987 released tremendous creative energy in Cass Turnbull's life. "I guess people call it passion," says Cass. "I had no previous idea that I had such passion. In fact, I had previously been plagued by depression. That certainly evaporated once I took action to change the world for the better. Now I have exhaustion, but that's better than depression. About anything is," she says and laughs.

As PlantAmnesty began to gather momentum, Cass continued building her private landscaping business. In the early years, Cass was working about thirty hours per week for PlantAmnesty, earning about $11,000 per year primarily for her teaching efforts. So, her landscaping/maintenance business is crucial financial support. "People kept saying, 'Do what you love, and the money will come' . . . but they were wrong. I got the fame, but the fortune did not come. It's fun, but the lack of money was a source of marital strain and personal sacrifice."

Finally in 2000, Cass decided to step out as executive director, turn the organization over to the nonprofit board, and focus on her own business, book writing, and personal life. Some good things came out of her four-year sabbatical—the organization got moved out of her modest home and garage, and she got her book written. In 2004, Sasquatch Books published *Cass Turnbull's Guide to Pruning: What, When, Where and How to Prune for a More Beautiful Garden*. But in general Cass felt like the sabbatical was ultimately unsustainable for PlantAmnesty.

"When I came back to the organization, they had run through $30,000 worth of savings, the board was turning over every twelve months, and they nearly lost their focus on the negative!" she quips. "It certainly was a hard time for me. People who originally adored me and would have followed me off a cliff suddenly wouldn't agree with me about anything. I became seen as everything that was wrong with the organization."

During one of the years of her sabbatical, Cass was completely out of the organization except for her teaching responsibilities—she didn't even write the newsletter. She described it as a time of huge battles and bizarre moments. "Thank God my husband was here to listen to me. I had some extra time, so I went back to having parties with friends, but I'd changed. I'd gotten hooked on the gratification of doing good. Parties are fine in college, but at some time you have to grow up." So, back into the organization she came—but not without a fight. When she ran for the board in 2004, she won by one vote in an election she described as similar to a scene from the television show *Survivor*.

"It was funny, though, the second I got back in, some giant gear shifted, and all of a sudden the organization got in line moving in the same direction again." In 2005, the first year Cass was back as president of the board, the organization earned $80,200 primarily from dues, referral service fees, donations, and the annual fall plant sale. Expenses that year, which included maintaining the office staff established during Cass's absence, amounted to $78,000. Overall 2005 posted a profit, the first one in five years.

On that budget they accomplished the following:

- They raised the awareness of poor pruning and maintenance through twenty-three media events that reached 6,815,000 people, including public service

Photo Courtesy: PlantAmnesty

⌃ *PlantAmnesty and Cass study a West Seattle cherry tree for possible inclusion in the Heritage Tree Program.*

announcements, newspaper articles, advocacy events, and a "shear madness" campaign.

- They provided solutions through classes and publications via the popular master pruner seminar series, the fruit-tree-pruning field day, and many slide shows given to garden clubs, professional organizations, and community groups. In addition to their existing literature, they published new pruning articles, answered eight hundred questions on the pruning hotline, staffed a booth at major city events, and made 2,263 matches from their referral service.

No, the referrals are not about dating but about matching tree owners with reputable and certified arborists who carry the Cass Turnbull seal of approval in how they'll touch a tree.

PlantAmnesty engenders respect for plants through the Heritage Tree Program for Seattle, media events, and a strong membership program. The program includes newsletters, an Adopt-a-Plant program, and a cyber library with hundreds of articles.

"If you stick up for the seemingly least important thing in the world—maintaining trees and shrubs—you can't expect to get a lot of money. My job is to get people to care about what they don't care about. The need to see trees as valuable—I don't mean in the commercial forestry sense but to value what's right in front of where we live and work and walk around—has been an issue in urban development forever." Cass knows some of this history.

"In the early 1900s, John Davey led a campaign to end tree butchery. He ran a campaign against tree topping and wrote a book called *The Tree Doctor*. He was an orator, had a radio program called the *Davey Hour*, and traveled the country. We're actually a lot alike except he was stiff and religious and told his crews to stop smoking and start exercising, and that's certainly not me. I'm more likely to say, 'Let's go to a hot tub and party.'" John Davey was a hardworking English immigrant who landed in this country in 1873 and launched his tree-man career by accepting a job returning the overgrown Standing Rock Cemetery in Kent, Ohio, to its former beauty. In 1901, assuming a debt of $7,000, John Davey published *The Tree Doctor*—a milestone in the science of tree preservation. By 1909, he and his son, Martin, founded the Davey Tree Expert Company and established a formal school where men were scientifically trained to perform tree surgery work.

Cass also explains, "The first time global warming was introduced was when the rain forest losses were being brought to light. There was even a tree on the front cover of *Newsweek* magazine. It was back when TreePeople planted a million trees, but then trees fell out of favor again." Trees are clearly "back in favor" again. That won't hurt PlantAmnesty's fund-raising ability. And the group certainly can take credit in the Seattle area for keeping trees in the public mind-set.

"My whole thing has been to get people to care about trees and plants by making a big fuss about them and giving them a dollar value. Let me offer an analogy. After all, I am the queen of the blunt analogy. Let's take modern art. I don't get it. But I wouldn't throw a Picasso on a fire. It's worth money, and I'd look like an idiot in front of my friends who do know about art. So, because we've made such a fuss about bad pruning and because we've assigned dollar values to trees, most people in this city are more careful of the pruning that happens in their yards. If you know there are people out there who would go dingy if you treated your plant with disrespect, you might not understand why, but you won't do it—especially if you think you might show up on the evening news."

Cass Turnbull and PlantAmnesty are forces to be reckoned with. They have laid down a template of public education that is working for trees and shrubs. Entering the city of Seattle, a person is immediately struck by the abundance of greenery. Large trees here are better cared for than before the advent of PlantAmnesty. If Cass has her way in the years ahead, the radius of good tree-care influence will spread to smaller trees and shrubs and out beyond the city to surrounding areas. And because so much good information is now available on the Web through PlantAmnesty and other sites, hopefully other latent activist pruners will wake up and take on public education in hundreds of cities and towns across the continent, ensuring a landscape of happy trees.

Seed cones of Pinus longaeva, the bristlecone pine. »

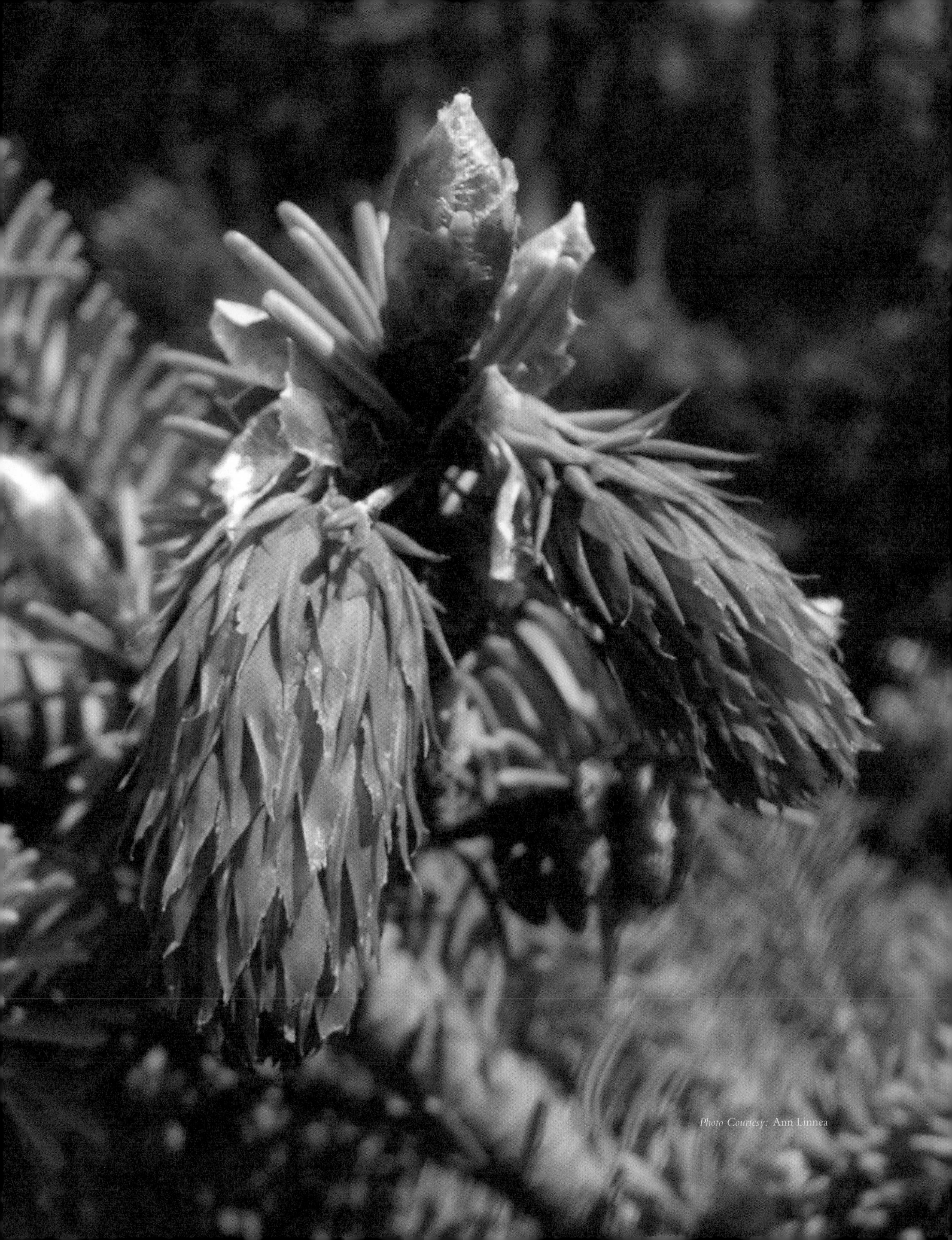

Photo Courtesy: Ann Linnea

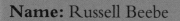

Name: Russell Beebe

Occupation: Carver, maker, and restorer of furniture

Point of Wisdom: The basic understanding of working with a tree is that you must work with the grain and let *it* tell you what to do. The key is letting go of preconceived notions.

CHAPTER 13

The Native Carver

Russell Beebe

Everyone is born an artist. Some of us learn a craft so we can express ourselves as artists.

—RUSSELL BEEBE

HIS STRONG, SURE HANDS show the weathering of a lifetime. He smiles and shrugs his shoulders when pointing out the loss of a fingertip in a planing accident years ago. It's the mark of his trade, the mark of a tree: fingertip, branch tip, it's all the same to this soft-spoken carver. He carries no extra weight—a tree doesn't get fat. His face is as grained as the lines of wood he follows with hand and chisel. There's a sense his life has been shaped by wind and sun, that he has endured cycles of drought and rain just as the trees around him. His hair is graying, a mossy shag on this head. His eyes are blue, and he sports a mustache—nothing calls attention to himself. The unassuming presence of indigenous people radiates out of stillness and his story.

The loading doors to Aaronson's Woodworking Shop near Ashland, Oregon, are open wide to the early autumn sun. The day brings warmth onto the concrete floor while Russell Beebe drinks coffee from a thermos cup, nurses along one of three hand-rolled American Spirit cigarettes he allows himself daily, and paces thoughtfully around the base of a partially carved seven-foot walnut stump. To watch Russell Beebe chisel tiny shaves of wood from the life-size carving emerging from the base of this tree is to watch creation come to life in his hands. "Carving gives me the most joy of all the woodworking I do because it is the most difficult," Russell comments. "In the U.S. we have nearly lost the art of carving because we are so mechanized and demand such quick results. The only way to carve wood is to listen to what the grain is telling you inch by inch and to move slowly—no big chunks, no quick decisions. Otherwise you're in a fight instead of a process."

≈ *Russell at work.*

"My work is one of a kind, so I must carve with chisels, slowly approaching the desired shape," Russell continues. "Texture is important. I want people to touch my sculptures, and everyone loves a chisel-cut finish. But when I want a rough background, I use a chain saw here and there. In carving an image, I will close my eyes and run my hands over the carving and continue working on it until it feels like what I am trying to carve."

Russell has traveled far and jumped from culture to culture to study with master woodworkers. He has mentored with Native elder drum makers. And he has done whatever jobs are necessary to keep himself housed and fed and his hands close to wood. "Carving is starving," he says, and smiles again. Now in his sixties, he finds himself in the position of carving statues of public significance. His most recent work, *We Are Here* (*Eme'tekliyikikh*)—also known as the Gateway Alder—is a twenty-foot memorial sculpture-carved alder trunk that stands at the entrance to downtown Ashland, Oregon, home of the acclaimed Shakespeare Festival. Commissioned to carve the Gateway Alder in 2004 as a gift to the community from a local businessman, Russell stepped into considerable controversy and worked with renowned Takelma elder Agnes Baker Pilgrim to help people understand that the statue was more than a piece of art—it was a healing, ceremonial pole for Native people that acknowledges their historical and remaining presence in the Rogue Valley.

When not involved in his own carving projects, Russell is the resident master in a local woodworking shop. "I get involved in a lot of wild stuff here," he says. "I restore old furniture, fix anything that goes wrong, do all hand work, and must understand traditional and modern technology. One project involved repairing a three-hundred-year-old concert zither from Saudi Arabia. And once I even helped a lady restore a fine piece of Meissen porcelain." It is the mix in his own bloodlines and his life experience growing up in the Great Lakes area that allow him to practice such a broad range of craft and artistry. Practical knowledge was fairly common there—kids tried to fix anything broken and were enthusiastic builders.

Early History

Not surprisingly, Russell liked to whittle from the time he first got hold of a pocketknife. Born in Baraboo, Wisconsin, during the early years of World War II, he spent lots of time in the woods. "I was born and raised rural," he remarks. "So I just knew about the different kinds of wood. I was always whittling—little boats, trains, and cars. One of the things I really wanted to be able to do was to carve a wooden muskie lure that was weighted in such a way that it would twirl when it went down through the ice—drawing a muskie in close enough to spear." Muskellunge, commonly called muskie, is a large uncommon member of the pike family of freshwater fish. These fish can attain lengths of two to five feet and can weigh over sixty-six pounds. Because of their size, they have few predators other than fishermen.

Russell knew of a few local carvers who could make the coveted muskie lure, and he tried to copy their models. It was part of his understanding that carving was art and the beginning of his study of shape, form, and the interplay of wood and outcome. This fascination with carving something that has a lovely shape, that people want to touch, and that "works" from all sides remains a hallmark of all his carvings today. He said he got this sensitivity from his mother, who was a Russian from Winnipeg, Canada. "She liked to paint murals on basswood cheese barrels and was really proud that the images and the stories went all the way around the barrel," smiles Russell. "My father came from the French Boebe in Normandy. I am one-eighth Ojibwe and a member of the Bear Clan."

For years Russell and his younger brother spent hours trying to catch their big fish. Eventually they got their muskie, though undersized—lured by a metal swim bait. By then Russell had moved on to carving other things. He has always enjoyed Native craft immensely. He learned to tan buckskin, work with birchbark, and do most other Ojibwe crafts. He was taught how to track a deer in deep snow, offer tobacco to the dying animal for the life it was giving, and make a temporary shelter. In the morning he was shown how to make a pair of snowshoes on the spot using the deer and what grew around and then pack the animal carcass out of the woods.

He attended high school in Milwaukee and then spent some time working as a commercial graphic artist there, further honing his skills. Like many young people, he was drawn to try a number of different professions. He applied his passion for line and flow to designing fiberglass auto bodies for sports cars in California for a few years. One of the designs he created was eventually bought by a man who in turn sold it to Datsun for the 240Z prototype. But working with fiberglass was very different from working with wood. He returned to natural materials. "Natural things like wood or stone have their own character. It's not really possible for us humans to just manipulate natural materials however we want. It's the relationship with the wood that I've always loved—how it teaches me what to do next.

"I remember taking the bark off of one tree I was going to carve," says Russell. "There was a big hole underneath, so I plugged it. I was doing my left-brain thing, thinking I had plans for that piece of wood. But that's not what the tree was telling me, and until I listened, nothing good was going to come out of my work."

And just as Russell is dedicated to listening to what the wood is telling him, he also has respect for learning the proper way to do things. "I was doing pretty furniture, but I knew I wasn't totally getting it right," gestures Russell with one of the carving tools he keeps close at hand. "One day I was in the Siam West store in Portland, and I saw the quality of woodworking I wanted to do. The oriental woodwork in that store was absolutely exquisite. So, I did some research, and in 1982, I traveled to Bangkok to learn woodworking and cabinetry from Jeng Yee, one of the seven master woodworkers in the world." Russell wrote about his learning experience in *Fine Woodworking* magazine (November/December, 1982, no. 37, "A Chinese Woodworker, Looking over Jeng Yee's Ancient Shoulder").

Returning to Bangkok in 1986, Russell availed himself of everything he could learn. He enthusiastically studied the quality and intricacy of Chinese woodwork. "I saw an eight-hundred-year-old rosewood bed. Using a Chinese process called 'glueless joinery' that massive bed was held together with only hidden locking joints. There wasn't one drop of glue in that bed! It looked brand new, and you could not figure out how to take it apart unless you knew the one central joint that everything hinged on." Jeng Yee died in 1987, and Russell has incorporated the master's sensibility into the very different style of his own work.

Developing His Native Roots

The 1980s were a time of tremendous growth and learning for Russell. First, he connected with the masters of the Far East, and then he connected with his Native American roots. When Wallace Black Elk, traditional Lakota elder and spiritual interpreter, traveled to Oregon, he was drawn to Russell's work. "He had lots of people wanting him to speak here and there, but he loved to just hang out and work with his hands doing traditional crafts," Russell says.

Raised on the Rosebud Reservation in South Dakota, Black Elk had been trained since childhood in the sacred traditions of his people. A direct descendant of Nicholas Black Elk, whose wisdom was shared in the book *Black Elk Speaks*, Black Elk was in his mid-sixties when he met the mid-forties carver and invited him to pay attention to the Native sources of his spirit and work. Russell found himself being mentored by Black Elk in the making of drums. "The first drum I coopered out of rosewood because it makes the best sound, and then we used Holstein for the heads. This was my first drum, and it was made for Wallace. He just kept showing up and tutoring me. I learned that an Ojibwe drum must always travel east, that it must be created within community, that it must hang suspended, and that it must be a living drum so it has feathers on it—not for decoration, but to tell a story."

"Black Elk became a guide and a friend. He got my enthusiasm up, and he pushed me home." At the time, Black Elk was one of the principal spiritual advisors to the American Indian Movement (AIM), located in Minneapolis, Minnesota. Spirituality and connectedness between all Indian people are central goals of AIM. Formed in 1968, the organization has taken numerous social justice actions to battle ongoing suppression of the civil rights of Native peoples. In the midst of sometimes fractious activism, Wallace Black Elk carried a message of peace both nationally and internationally to all who would listen. Russell listened and headed back to his Midwestern Ojibwe roots.

"The AIM guys all hung out in Minneapolis," explains Russell. "It was there I met my friend 'Porky' White." Walter Gahgoonse White was born in 1919 at the Federal Dam, Leech Lake Ojibwe-Anishinaabe reservation. He was nicknamed Porky because Gahgoonse meant "Little Porcupine." His mother's sugar bush camp provided Porky with important skills and knowledge of indigenous traditions that he shared with Russell in the years of their friendship. Russell was especially touched by the stories of Ininatig, the man tree, who had taught his Anishinaabe people about making sugar from the first rising of the sap in the maples. He taught the people so they wouldn't starve in that vulnerable time in the north when winter is alternately releasing and reestablishing its grip on the land. "There's something magical about being in the sugar bush camp in those cold days of late winter and seeing that sap flow, being around the fires helping to boil down the sap to syrup," says Russell.

The man who did not grow up on a reservation was reconnecting to his earth-based upbringing and expanding on the inherent knowledge his people carry about trees, forests, and all living things. "Hanging around Indian people, I discovered a lot comes through the bloodline—especially following intuition. Porky was key in continuing my education, and he introduced me into ceremony," explains Russell.

"After some time in Minneapolis, Porky and I went across the St. Croix River into Wisconsin, and he introduced me around. I carried with me a second large drum called the 'big drum' [*chi de wegun*] built in the traditional way. People there saw that the drum I had made was done well. They found everything right about the drum, and so they began to accept me. At Mole Lake, Wisconsin, I was initiated into Three Fires Midewewin. Following the four-day ceremony, my drum was given to the Sokaogan band of the Ojibwe. This was a huge turning point for me. I became a maker of Three Fires Midewewin drums."

The Midewewin is a major religious ceremony of Anishinaabe peoples. It is their medicine society. Upon initiation into the Midewewin, Russell began to make conical

« *Russell beginning to reconnect with his Native American roots.*

water drums for ceremony. Water drums are considered the most sacred and rare of all drums and probably originated with the Ojibwe people. They are made with special wood from certain trees. The drums are filled with unique amounts of water to create particular sounds once covered with tanned deerskin. The drums are used only for sacred ceremony, and the limited public knowledge of the drums and ceremony is respected here. Walter Gahgoonse White died in 2001, and Wallace Black Elk died in 2004, and by then, Russell Beebe, Wabashkigamash (Storm Wave), was remembered to his people.

Carving Career Begins

The growth of a tree teaches us that things take time. A seed drops into a fertile crack next to a rock in the forest and a seedling manages to grow—in the north, perhaps two feet in its first decade. Miraculously, if conditions of weather, placement, and protection continue to be favorable, the sapling takes its place on the forest floor as one of the up-and-coming strong trees. One year a huge storm rages and fells a nearby large tree, creating an opening in the upper canopy. There is an opportunity for the young tree. Many decades pass, and gradually the tree begins a new role as one of the old growth in the forest. The life of a tree is not a pathway of shooting to prominence or a pathway that any other tree has exactly followed in exactly the same way. Following the opportunities that present themselves and survival among the ever-changing odds of nature is the pathway understood by Native peoples.

All of Russell's experience with master woodworkers in the Far East and with his own Ojibwe elders prepared him for opportunities that were to come his way. "About eight years ago I got my first commission as a carver," says Russell, who was previously identified locally as a fine furniture maker and builder. "A federal judge asked for a new door for his house out of solid mahogany. First, I made the door. Then he asked if I could carve it. It's not my preference to work with rare and endangered woods, but I took the job. After that carving, commissions started coming."

A local Ashland businessman and longtime patron of the arts, Lloyd Matthew Haines, wanted to make an art piece out of the dying white oak in his yard instead of cutting it up for firewood. He approached the local Siskiyou Woodcraft Guild, looking for an artist. Several prospective carvers responded. When Russell asked Haines what message he wanted the statue to convey, Haines knew he had the right person for the job. Haines's desire was for a sculpture that would honor the tree, the wildlife, and the Native peoples who once lived in the area. Scaffolding was placed around

Russell with completed statue of the white oak trunk. »

the remaining twenty-foot trunk of the tree, and Russell began to listen to what the tree wanted and what Haines had articulated. It was not uncommon to see the two of them up on the scaffolding in the early summer evenings discussing how the artwork was unfolding and how it related directly to their own lives.

In the summer of 2003, after three months of full-time work, Russell completed the *My Relatives* (*Akinaa Bimadezin*) statue. As the plaque states, he carved it "to honor this tree, the earth, and my relatives." Using both a chain saw and hand tools to bring forth the faces of the Native peoples of the Americas on the lower portion of the statue, he also deliberately chose to carve the Raven of Naas about three-quarters of the way up. In Northwest Native traditions, the raven is part of the creation myth—either as the creator or as integral to the creation. (Naas is a river in British Columbia that borders the Tlingit nation.) From the beginning Russell had envisioned the grand, three-dimensional eagle that crowns the statue.

Native art is rarely anatomically detailed. The shapes and figures are generally representative and invite story. "We are a humble people," explains Russell. "You won't see us putting detailed breasts on a woman or anything like that." Russell signed the statute with his spirit name, Wabashkigamash. Though he could not know it at the time, this statue was preparation for the next big project coming. After carefully treating it with Australian Timber Oil to keep fungus from growing, Russell returned to furniture building and other projects and teaching in his own shop.

We Are Here Statue

When Lloyd Matthew Haines decided to build a new eight-thousand-square-foot multiple-use office building in downtown Ashland at the corner of Main Street and Lithia Way, he was consciously deciding to upgrade that corner of Main Street. "Previously there had been a mortuary, a veterinarian clinic, and a bar on that corner—not in that order," said Haines. "Drugs were routinely exchanged in the area, and lots of drinking was happening there." Haines's vision was to create an aesthetic building for offices, a restaurant, and residential space—an inviting impression to downtown from the north, a main entrance point from Interstate 5.

Three architects worked with Haines to design a beautiful, welcoming space. Everything was on track for approval with the city until it became clear that an old alder tree that had shaded the deck of the Ashland Creek Bar and Grill for years would have to be cut down to make room for the building. At one point Haines even attempted to move the building twenty feet, but ultimately it became obvious the tree would have to be removed if the building were to be erected. Haines decided to commission Russell to carve a statue out of the tree as Russell had done on his property and donate it to the city—a commemoration to the Shasta and Takelma tribes of the region.

Ashland is a city that loves its trees as much as it loves Shakespeare. Built into the foothills of the Siskiyou and Cascade mountains about fifteen miles north of the California border, the city has preserved and planted a diverse urban forest of native madrone, oak, and fir trees. Alder is also a native local tree and is most commonly found in disturbed areas such as clear-cuts or in open spaces along rivers and wetlands. As a species, its role is to restore nitrogen to the soil and to grow a quick canopy of shade under which the conifers can reestablish themselves for the long haul. The usual life cycle of alder is forty to sixty years from seedling to blowdown. The contested alder had been growing near Ashland Creek for over fifty-two years—the elder tree in the area, both magnificent and vulnerable. Once a few people got wind of the proposed tree cutting, the controversy was on.

Some of the objections came from people who quoted a Celtic saying, "He who cuts the alder, curses the village." Others objected strictly out of a love for trees. Still others objected to the general idea of change. Ironically, in the articulation of the Celtic Tree Oracle, an alder's role is to provide "spiritual protection in disputes." So while the twenty-first-century inhabitants of Ashland entered the fray of controversy, ancient energies were at work from both European and indigenous roots.

Russell got caught in the middle of the controversy. "I was called a tree butcher at one point," he says. "I had to appear before the Ashland City Council along with Mr. Haines. We asked Agnes Baker Pilgrim to come with us."

Takelma elder, Agnes Baker Pilgrim, blessing the living alder. »

Agnes Baker Pilgrim is a nationally and internationally recognized spiritual elder and teacher. She is the oldest living Takelma among the Native peoples of southern Oregon and has revived her people's Sacred Salmon Ceremony. In 2004, she was called into the International Council of the Thirteen Original Grandmothers who travel around the world to raise consciousness about the dire state of the earth's well-being through prayer and support for local projects. She calls herself a "Voice for the Voiceless," and in the true manner of an elder, she can speak volumes of truth with a few words. When Agnes rose to speak before the city council, she said, "My people have lived here in the Rogue Valley for thousands of years. The only evidence that we were here at all is a road outside town named 'Dead Indian Road.' It is time that there be a memorial to my people." The combination of her presence and the truth of her speaking shifted the entire tenor of the meeting. The building was approved, and on Election Day 2004, the alder was cut.

Throughout the autumn, the tree was watched carefully by many—including Russell. He spent hours sitting with the living tree, seeking direction for the design that was to emerge. Slowly a drawing of the statue became clear to him. Prior to the tree being cut, Agnes led a blessing ceremony for it. "We prayed to the tree, walking all around it. We smudged it and talked to it and promised we were going to make it into something beautiful for perpetuity," she said.

⨟ *Russell's sketches, tools, and the scaffolding around the We Are Here statue.*

On the actual day of the tree felling, many people, including Russell, came to watch. It was a noisy process, a big contrast to the contemplative moments Russell, Agnes, and others had spent with the elder alder. Chain saws and wood chippers roared constantly. When the tree finally came down, it was lowered onto the street too heavily. The large branch that was to have served as the basis for Russell's original design shattered on impact. "When that branch broke, I had to start all over," Russell sighs. The tree was delivered to Russell's home and secured between four trees so that scaffolding could be erected and the artist's work begun.

Carving a tree that has so recently been cut requires careful accommodation between what the artist envisions and what the wood itself is doing. The trunk of a tree is a magnificent conductor of liquefied nutrients—from the roots to the branches with their leaves or needles and back from the leaves or branches to the whole tree. According to Russell, the trunk wood in most species dries at the rate of about an inch a year. Even though cut, the trunk of a tree keeps up its action as a cellular hose for a long time. "When I carve a tree, I imagine a barbershop pole," explains Russell. "I want to cut close to the center and then move back out to the edge to allow moisture from the core of the tree out to reduce the cracking. Totem poles were traditionally made from cedar because it is straight grained and rarely rots. Most are halved 90 percent up to reduce cracking. There is very little straight grain in an alder. The Gateway Alder taught me a lot."

Russell's charge was to honor the first nations of the local region—including the Shasta and the Takelma (Agnes's tribe). (Native peoples include all wildlife, trees, rocks, and flowers as nations.) "I put a lot of pressure on myself to do this really well. I sat up in that platform and agonized about how it should look—about what forms the tree was telling me should emerge." Both Agnes and Haines occasionally spent time on the scaffolding with Russell. Even though she was in her eighties, Agnes didn't hesitate to spend time perched next to Russell. She wanted to provide moral support and prayer, especially during some of the short, cold days of winter. "He has such a way of carving," Agnes said. "It's unbelievable that design came from one person's mind."

Sometime after the branch broke and Russell was wrestling with the new design, he "saw" the head of the deer emerging from the end of a broken limb on the upper part of the pole. The rest of the design flowed, and he could follow the teachings of the tree quite naturally. He was always planning to carve a woman. Several months into the project, Agnes came to visit. While Agnes was standing there, Russell's girlfriend challenged him to depict Agnes herself. How could he say no? Although it was

a tense moment at the time, ultimately everyone agreed it was a good thing. Russell asked Agnes to bring photos of herself as a young woman and samples of ceremonial jewelry she owned. A younger version of Grandmother Agnes is depicted in the statue arms raised in a gesture of prayer, bearing in her womb the coming generations. On the other side of the trunk Russell carved a Shasta man holding the next generation as a child on his shoulder. The togetherness of nations is shown through humans and animals—snake, salmon, bear, cougar, goose, raven, and deer.

It was important to Russell that the statue come alive from all angles. Working with the tree was key to what he did. "It was not just a block of wood that I could carve with whatever came to my mind," he explains. "This was a consecrated tree that had been taken with ceremony and would carry great ceremony. I had to trust the natural shape of the tree and the representations of the humans and animals that came forward to me."

Russell carved nearly every day for eighteen months. On September 30, 2006, the statue named *We Are Here* (*Eme'tekliyikikh*) was erected at the entrance to town. It was carefully placed onto its specially prepared base. Main Street was blocked off, and over a thousand people came to witness the arrival of a ceremonial Native pole in the middle of downtown. There are many stories of tremendous healing from that day.

"The statue became so much more than a work of art," said Haines. "On the day of the celebration, three or four of the most vocal opponents came and hugged me saying, 'Boy, our hearts have changed.' It is a memorial to the First Nations of this area, and it has become a site for tremendous peace and healing. I was completely overwhelmed with emotion when it finally went up."

"We kept our word to the tree about making it into something beautiful," said Agnes. "People have called from all over the world talking about the importance of creating a ceremonial pole in our city."

"It is always difficult for me to decide when to pack up my chisels and say, 'That's it, I'm done.' This was especially true on this piece," says Russell. "I had focused my life with this tree, completely on my self-imposed quest for perfection. When I finally reached completion and put my tools away, I sensed very strongly a release, that now the sculpture had a life of its own. I was surprised to feel then as a witness to its vitality. This new relationship became so clear during the dedication ceremony. When we began our tobacco offerings in prayer, people from all walks of life felt the urge to participate. They came by the hundreds asking for a pinch of tobacco and spent

« Detail of "coming generations" figure within the **We Are Here** *statue.*

private time close in praying for family, friends, community, and the world. I just felt so good to be among them. I'll never forget that moment."

"I am really honored to have done this work," continues Russell. "It's about as big an honor as you can get. By creating a sculpture, this tree and the remembrance of local Native peoples will now last forever." Forever is a long time. For now the sculpture is living its ceremonial promise.

"This started as a memorial and has become a prayer pole for my people. The tribes do ceremony with *Eme'tekliyikikh* at each of the equinoxes and solstices. Maintaining the Gateway Alder is a lifelong commitment for me, and this also needs to be a community project," explains Russell.

After several years in the Oregon weather, the statue began to darken and gather mold—even though it had been carefully finished with wood oil. Russell approached the city about their promise to maintain the statue. Bureaucracy must understand the power of the elements—wood is a very different kind of statue than bronze or stone. Even in a small city, the complexity of who is responsible for what can get tangled. Three years passed, no action was forthcoming, and the statue was clearly being damaged by both the drying and cracking power of the sun and the molding power of the damp, wet winters. Russell decided to take action.

With the volunteer assistance of a Skyjack, Russell was hoisted to the upper reaches of the statue in late summer 2009. It took him fifty hours to recarve most of the upper one-third and spots below and refinish the newly cleaned statue. Completing the work in 103°F heat, he became more determined than ever that the city keep its commitment to honor and maintain the statue. In his quiet, determined way, he brought Agnes Baker Pilgrim and Lloyd Matthew Haines into the effort. In early October 2009, in the presence of his two allies, Russell and the city signed an agreement for the lifelong maintenance of *We Are Here*. Russell has been designated to serve as the main consultant in preservation and maintenance but will not be reimbursed for his restoration efforts. This time there was a signature, not just a verbal agreement, and Russell has a copy of the contract.

Future Projects

Russell embraces the old tradition of elder teaching. "I want to be helpful and teach the next generations what I can," he says. "I cannot do my large works forever, so I need to pass my craft on to a younger strong person who can work with dedication.

Takelma elder, Agnes Baker Pilgrim, leading ceremony at the dedication of the **We Are Here** *statue, Ashland, Oregon, September 30, 2006.* »

The Native Carver 215

I'm hoping this person is my workmate Lucio Mejia, a strong Native American from Mexico."

At age sixty-seven, Russell is working on another public statue. The figures are emerging from a seven-foot-tall walnut stump. He has been commissioned to create a statue for Sanctuary One located at Double Oak Farm. The relatively new facility is "dedicated to rescuing domesticated animals, facilitating wellness and trauma recovery, and promoting environmental stewardship on an eco-friendly farm in Oregon's scenic Applegate Valley." Horses, goats, geese, chickens, and even a pig have been rescued. In turn the sanctuary offers programs and services for people, like those at the Southern Oregon Child Study and Treatment Program, which has a facility that promotes in-service learning and volunteering.

Statue size is not necessarily an indication of challenge. Although this carving is smaller than either the *We Are Here* or the *All My Relations* poles, Russell is again in challenging dialogue with a chunk of raw tree. Standing in front of the statue that is beginning to take shape in Aaronson's Woodworking Shop, Russell is his usual quiet, understated self. "If I waited for a twenty-inch-diameter stump to dry before starting to carve, a person like me might not be around to finish it. Like the Gateway Alder, this wood is fresh, but it's from a very different tree. When I got it from one of my Mexican American friends, it looked like a piece of broccoli with branches coming out everywhere. Every one of those branches comes from the middle of the tree and changes the grain. There are many lessons it teaches, as design is compromised anywhere a branch or rot emerges. This tree was dying, half live, half dead. If rotten wood remains, it must be impregnated for integrity. I use cyanoacrylate glue."

There are technical, as opposed to design, issues with every carving, and the wild walnut offers many challenges. A tree's trunk, like the trunk of our own bodies, is the stabilizing vertical portion that supports development of limbs. The human arm is supported by intense musculature to allow the trunk to support the arm's weight. A tree branch is supported by a tree's version of musculature in which the vascular tissue of the trunk moves into the branch, setting down its weight-bearing lignin. After the branch is gone, the woody musculature is forever inside the tree. When cut into planks, this support shows up decoratively—think knotty pine paneling. When the tree is offered to the hand of the carver, he has to work with every knot, follow the grain pattern, and "see" inside how the grain can accentuate the chosen image.

Moving in and out and around the walnut trunk, Russell works to "open up the center" so that moisture has a chance to escape thus minimizing cracking. When cracks do appear, he has to pound splinters of wood into them for stabilization. As he moves in and out from the center, he exposes both the sapwood (that part of the trunk that was still living and has more porous grain) and the heartwood (the denser core of the trunk that was dead but not rotten). The dark striations in the carving reflect the sapwood that is exposed at different depths of the carving. Letting the heartwood and the sapwood come through directs the design of the project as much as the locations of each of the multitudinous original branches.

This commission calls Russell to capture the spirit and essence of the mission of Sanctuary One—to restore the connection between humans and animals. He began by carving two mating red-tailed hawks. He wanted to portray both the wildness and the tenderness of their courtship. Moving around the sculpture—Russell's carving always tells a story in a circle—Russell has carved a cat, symbolic of the domestic animals that come to the sanctuary for care. The animal figures swirl around the androgynous face and torso of a person who relates to them in the Native spirit of "all my relations."

Relationship is the essence: between Russell and the tree; between the wood and the chisel; between the carver and the community; between the people and the valley; between the valley and the mountains. To draw his design out of the wood, Russell must both surrender and lead with his profound practice of interconnectedness with all things. Small cut by small cut in great patience he keeps the lineage of carving alive, and he keeps the traditions of his people a vibrant part of the artistic landscape. He is tapped into the spiritual roots of his Native heritage—his heart connected to the land, and the land connected to his heart.

Photo Courtesy: Marie Mannatt

Name: Ann Linnea

Occupation: Writer, educator, and wilderness guide

Point of Wisdom: People, like trees, can transform energy from the earth. Trees, like people, can be a source of helpful wisdom in our lives.

CHAPTER 14

The Botanist Grandmother

Ann Linnea

The wild channel of communication

between trees and me that had been opened on my journey

around the lake refused to close.

—ANN LINNEA

ALL OF MY LIFE I HAVE loved wild places—those nooks and crannies of nature furthest from the sights and sounds of civilization. The higher and more rugged the mountain, the larger and more remote the body of water, the deeper the forest, the more I feel at home. I have devoted my life to learning how to live in and with wilderness. And I have devoted my life to taking people into the wild with me, guiding them safely through nature adventures that change their way of being in the mechanized world.

I am quiet by nature, and nature has quieted me even further so that I can observe without interfering with the patterns of creatures and the whispers of trees. As a slightly built, now graying woman, I do not strike a pose of ferocity, but on the subject of respect for and relationship to nature, I am fierce. I share the story of my own love affair with trees and my particular modes of relating to them as a clarion invitation.

Early Years

Growing up in Austin, Minnesota, I delighted in climbing trees, sleeping in their shade, and jumping into piles of leaves. Trees were companions, guardians, even play-mates. They were safe. They held me even in their highest branches. My mother didn't worry when I asked if I could go over to Kim's house and play in the trees by Turtle Creek. The safety of a small Midwestern town in the 1950s gave me freedom to forge a strong attachment with the great elms, oaks, and maples that graced our town.

In the 1980s as a young mother, I looked at trees with a more pragmatic eye. My children and I established a small maple syrup operation in the snowy March woods across the street from our home in Duluth, Minnesota. We installed about a half dozen metal taps into assorted available maple species in a small patch of woods behind one of the local university dorms. Every day for about a two-week period in March when nighttime temperatures dropped below freezing and sunny daytime temperatures rose above 38–40°F, the sap would flow. We three donned our fleece-lined Sorel boots, down jackets, and wool hats and mittens and trudged through the crusty spring snow to find our taps. As grade-schoolers, Brian and Sally loved removing the gallon ice cream buckets we left hanging from the spouts and pouring the clear liquid into our big five-gallon collection bucket. The next step was to put their mouths up to the tree taps to taste the sugary liquid. After replacing the ice cream buckets, we would carefully balance two partially filled five-gallon buckets of sap on plastic sleds and pull them out of the woods across our usually snowy street and then up the hill to our front porch. There we poured our collection into a heavy metal pot and moved the liquid onto our wood stove to begin the several-day process of slow evaporation that ultimately produced the thick, sweet syrup that is a trademark product of the northern sugar maple.

We heated our northern Minnesota home primarily with wood, so I quickly learned to appreciate the different trees in our forests for their ability to provide heat. Birch and pine were both locally abundant woods and when properly dried have a heating value that compares favorably to coal, oil, and natural gas—and it is a renew-able resource. The ability of a particular type of wood to provide heat is measured in

⚐ *As a gradeschooler, Ann's son, Brian, loved to sample sap directly from the maple trees.*

British thermal units (Btus). Charts and tables have been figured out for the heating ability of nearly all types of trees. Birch measures well on those charts in terms of Btus produced per cord of dried wood, as does pine. (A cord is a stack of wood four feet by four feet by eight feet.) However, pine contains more pitch and therefore can more readily produce dangerous creosote buildups in chimneys. Birch eventually became our wood of choice. Our family loved the warmth of the little standing airtight stove in the living room. On bitter cold winter nights, everyone gathered around the stove to read, do homework, or play games. And both of my children learned the fine art of building a fire inside a wood stove and keeping it going. They learned to open the damper, create a tipi of smaller twigs surrounded by larger twigs, and slowly add larger pieces of wood until good coals were set in place. When they stoked the fire, they were careful to open the stove door slowly to keep smoke from back drafting into the house. To this day I would trust either of them, now young adults in Los Angeles and Denver, to know how to find properly dried wood, how to lay a fire indoors or outside, and how to keep it safely going. As a young mother, Sally is passing on that same sensitivity to my dear grandson.

Spending time together in nature is something I did with my children from the time they were tiny. We took long camping, hiking, and canoe trips together and en-

Photo Courtesy: Joe Villarreal

222 Keepers of the Trees

joyed skiing and skating in the winters. Trees were an everyday part of our existence in the north woods from sap gathering to fire building to shear enjoyment. Duluth provided us a good life—a city at the edge of great wildness both in the woods and on the water.

At age forty-three, in the summer of 1992, I set off on a ten-week wilderness kayaking trip to explore the forested perimeter of the lake we lived on—Lake Superior. I paddled off from the shore of my home city of Duluth in a well-stocked seventeen-foot sea kayak accompanied by my friend and teaching colleague, Paul Treuer. A lifelong athlete, I had trained for four years to prepare myself for this journey. It took us sixty-five days to paddle 1,200 miles in one of the coldest, wettest summers ever recorded. We camped and kayaked our way around the full circumference of the world's largest inland sea. My journey was filled with hardship, danger, beauty, and mystery as this remote and wild lake challenged every assumption I had brought onto the water and stretched me beyond my limits. Focused intently on the danger of these waters, what I didn't expect as I paddled away from home and family on that cold June morning was how often the trees that lined the shore would serve as my refuge and teachers.

Near the end of the first week of paddling, Paul and I parted company for a week so he could complete some work details. I proceeded along alone until he could join me. A particularly difficult day revealed the immensity of the challenge I had undertaken. Though the sun was out, the temperature at noon on this June day was still only 42°F. (Lake Superior is infamous for the cold temperatures of its water, which, of course, affects the temperature of the air near its shoreline.) After about four hours of morning paddling, I stopped for lunch on a tiny cobblestone beach. This was an easy amount of paddling for me—well within my conditioning level.

I enjoyed a sandwich and part of a thermos of hot pea soup and then pushed off in search of a larger beach on which to camp. My marine-band VHF radio indicated rain and wind would move into the area by late afternoon. I noticed high, thin cloudiness but assessed that I had a few hours to find a suitable spot. However, the lake surprised me. Within fifteen minutes of sliding my little boat back into the icy waters, the wind began to pick up—changing the water's surface from flat calm to a one-foot chop. I watched the cloud patterns with a wary eye. Within half an hour, the large fog bank that had been sitting some distance offshore enveloped me. A steady mist began falling. The wind, clouds, mist, and fog were all danger signs. I needed to immediately find a safe place to land, set up my tent, and get off the lake. Cliffs, large rocks, and irregular cobbles make up much of the North Shore of Lake Superior. To land my kayak in wavy conditions, I needed a shoreline of sand or small stones.

Seeking the Safety of Shore

A kayak is an intimate craft. It fits securely around the paddler's body—more like wearing a boat than riding in one. The walls of my plastic Aquaterra sea kayak were about a half-inch thick: One-half of an inch was all that separated me from Lake Superior's 36°F cold. Cold water below me, cold water falling from the sky. Cold water seeping through my paddling jacket into the neoprene wet suit, where my body was barely capable of warming it. Despite the exertion of paddling, my reservoir of warmth and strength was being steadily depleted.

I propelled the yellow kayak through the fog, rain, and waves on the strength of my arm muscles, seeking to find a suitable beach that would enable me to land in building surf. Questions ate at my mind's rigid control, but I would not succumb and kept ordering my body, "Push on. Push on."

After nearly two hours of steady paddling, a strip of black sand appeared. It was not a mirage. I paddled closer. The waves were rolling in three to four feet high and crashing on the small slant of beach. I hovered just beyond the edge of the breaking surf, back paddling to hold my boat in place while I watched the pattern of waves. One careless move and I would be dumped into the ice water, pushed over the razor-thin line that separated me from hypothermia. Four large ones. Three small ones. Five large ones. Two small ones. Four large ones. With all my remaining strength, I pushed my boat and myself forward to ride in on the small waves.

Thump, the boat hit the sand. "Yes!" I shouted out loud. "I did it!" Quickly, I pulled my spray skirt free, jumped out, and dragged the boat up and away from the waves. I removed my paddling mitts and opened the front hatch to pull out my tarp and tent to set up a campsite. My fingers were so numb I could barely tie the corners of the tarp to trees, but once it was up, I grabbed my bag of dry clothes and sat down on the ground under the tarp's shelter.

Bending over from my sitting position on the ground, I reached to remove my kayaking booties. Suddenly a muscle spasm started in my side and spread to my stomach and thighs with the speed of a lightening bolt. I collapsed onto my side, frozen in a fetal position of excruciating pain—my cold, overtaxed muscles locked in

Paddling her kayak through the fog, rain, and wind, Ann was searching earnestly for a safe camping spot. »

helplessness. I couldn't move; it hurt to breathe. I was totally, utterly helpless. My body had never cramped to this extent before. My mind could no longer control and solve the problem. I needed help now.

I looked at the young aspen trees whose slender trunks were holding my tarp. The branches hanging below the edges of the tarp seemed to be reaching toward me, gesturing their willingness to assist, but, of course, they couldn't move toward me. "What should I do?" I pleaded with my mind. "Please help me. What should I do?"
Breathe.

The first breath was so painful, I dared not try again. But there were the trees. They were alive. Breathing in their own way. Another breath went in. The pain was deep, but I could breathe on the surface above it. I had the reassuring thought, "I am not going to die."

With each breath slightly deepening, my overtaxed muscles slowly eased their grip on my life. Finally I could relax my stomach and slowly straighten out my legs. I lay on my back under the tarp and looked at the trees. "Thank you," I whispered aloud.
Keep breathing.

I lay on the cold, wet ground for what seemed a long while, afraid to sit up for fear the cramp would return. But I was calm. Among these aspens, I was a lone human, but I was not utterly alone.

Photo Courtesy: Ann Linnea

Move slowly.

I rolled onto my side and eased myself into a sitting position. No cramp. First I unzipped my life jacket, and then I pulled my paddling jacket over my head. I was aware that a sudden movement could send me back into spasms and that the violent shaking of my hands indicated I was progressing through stages of hypothermia. Finally I was standing up, completely undressed. As quickly as my clumsy cold fingers allowed, I pulled on long underwear, fleece, rain gear, boots, and warm socks. Finally I was dressed and dry.

"Food. I must eat something." It comforted me to speak aloud, to test the clarity of my thoughts, to converse with the listening trees. I walked over to the kayak, which rested just beyond the grasp of the lake. I struggled to unclasp the back hatch cover so I could pull out my stove. Even though earlier I had opened the front hatch to retrieve my tent, sleeping bag, and clothes, my fingers no longer had the coordination or strength to release the plastic buckles on the back hatch.

I saw on the kayak seat the thermos containing some remaining pea soup, so I reached for it and managed to unscrew the lid and pour. Wonderful warmth! Soothing liquid.

"Tent. I must get warm. Tie the boat, turn it over, and get into the tent."

I managed to get the tent set up under the tarp. Within minutes I had crawled into my tent, zipped it closed, removed my boots and rain gear, and climbed with my wool socks, long underwear, wind pants, wool sweater, wool cap, and gray fleece jacket into the womb of my sleeping bag. I lay on my back staring up four feet at the dry roof. Listening to the rain and wind beating on the tent and tarp, a ruthless stalker seemed to ask, "What if that muscle spasm had come while you were paddling?"

I cringed, pulled the sleeping bag over my head, rolled my body onto its side, and tucked my knees up to my chest to protect my soft belly. "Drowned," I whispered. "I would have upset the delicate balance I used to hold the boat upright. And I would have gone over in a totally helpless, cramped position." The wind rustled the aspen trees against the tarp—a whispering, shushing sound. And then I slept.

Relationships with Trees

In that desperate moment I learned that trees, too, are relational. From my college botany years, I knew that aspen trees related to each other by growing in cloned clumps—many trunks connected to one gigantic root system. Yet, even after

As Ann continued paddling around Lake Superior, she often found comfort sitting on the shore with trees. »

Photo Courtesy: Ann Linnea

The Botanist Grandmother 227

hundreds of wilderness experiences, I had never considered that *I* might find my-self in a "relationship" with a tree. When I awoke the next morning, I felt genuinely companioned by the small grove of young aspens protecting me from the offshore winds still pushing waves onto my beach. I broke down camp and skillfully eased my loaded kayak back into the lake. From that day on, during the remaining nine weeks of the trip and in the demanding months of reentry to my life, I often spent time with trees—gathering calmness and wisdom beyond the realm of language.

About halfway around the lake, I camped among a small grove of scrubby subal-pine firs about thirty feet up a steep blueberry-covered slope on the Canadian shore. After landing my kayak, I took two trips to carry the waterproof stuff sacks containing my tent, sleeping bag, toiletries, and clothing up to the little cliff-top ledge. I left my food and cooking gear nearer the water's edge where the kayak was safely tied. I was feeling lonely. My five-foot, eight-inch slender frame stood shoulder to shoulder with the little firs eking out an existence on the thin layer of topsoil. The trees and I were both tenacious and wind sculpted. Sitting among them, I began to experience the depth of connection offered me by the natural world. I belonged in this landscape. I had a place with nature. The trees and I stood side by side as companions.

Relationship is commonly defined as a connection, association, or involvement. We humans generally view relationship to mean a connection with other humans or animals—beings that look back at us. I believe the possibility for relationship also exists between humans and trees.

In the years after the lake, resting quietly in the companionship of trees has be-come the primary spiritual practice of my life. When I first discovered this practice, it was lifesaving; now it is soul saving. My first year back from such a profound journey was as challenging as the journey itself. I had so totally left the routines of my life that I did not know how to find my way back in. I was a misfit even at the most basic physical level. My sense of hearing had become so acute that the family agreed to set the phone ringer to mute. My sense of smell was so keen that I couldn't walk down the street without smelling neighbors' backyard garbage cans. I had gotten as close to becoming primitive as a white, North American, middle-class woman could become at the end of the twentieth century.

While my husband, children, and friends waited for me to return to "normal," the trees coaxed my learning at a deeper level. Sometimes when I was passing a particular tree, a shudder would pass through my body—not a cramp, more like a power surge, a jolt of energy that seemed transmitted from the tree to me. The wild channel of communication between trees and me that had been opened on my journey around the lake refused to close. I had no name for these experiences, didn't know what to do with them, and didn't feel safe speaking about them.

At the time, a friend had enrolled in the Healing Touch Program. As I listened to her speak about her work connecting with the healing energy of the universe, the thought occurred to me that I might have become sensitive to the energy of trees. At the time, the study of "energetics" was completely new to me, but it provided a bridge between my recent experiences and the security of my intellect.

Started in 1989 as a continuing education program for nurses and other health professionals, Healing Touch teaches people to use their individual energy fields (the low-voltage "electricity" that emanates off all living things) to support their natural ability to serve as healers to their patients and themselves. The theory is that people in sickness have weakened energetic fields, and a Healing Touch practitioner serves as a conduit drawing on universal energy to replenish the patient's energy. Over the next several years as I took Healing Touch classes, I began to notice how the energetic field of one living thing could affect the energetic field of another living thing. As my fascination grew, I studied yoga and Chi Gung to gain better awareness of my own energetic field. I had spent forty years gathering the physical skills to explore my beloved outer world—kayaking, hiking, backpacking, skiing, and biking. Now past the half-century mark in age, I was shifting the focus of my physical skill building to those things that helped me better explore my internal world.

Early Plant Connections

Both of my parents, Astrid and Frank Brown, love the natural world and instilled in me a sense of confidence, skill, and respect for learning more about the earth. When I went off to college, I was eager to study natural things. As a botany major at Iowa State University in the late 1960s, two professors, Dr. George Knaphus and Dr. Lois Tiffany, augmented my love of nature with sound academic knowledge. They laid in place a template of curiosity and care that has guided my professional life as a writer and teacher of outdoor skills and knowledge.

Dr. K had an open office where students hung out amid his extensive library and towers of unfiled papers. Wearing a suit and tie with his gray, balding head constantly looking around the auditorium, he routinely captivated college students enrolled in Botany 101 with his lecture about photosynthesis. By running from one end of the chalkboard to the other gesturing wildly about plants taking in carbon dioxide and producing oxygen for us to breathe, he created enough drama to engage even the most disinterested student. Invariably he would call on some slouched down, well-built guy in the back row taking this as his one required science course. "What is the secret to life?" Dr. K would ask him.

"Chlorophyll?" the fellow would answer hesitantly.

"That's right! Chlorophyll is going to help you throw that winning touchdown pass on Saturday," Dr. K would respond to gales of laughter.

Dr. Tiffany was a more subdued lecturer with a neat office and a quiet, dedicated presence. A small, poised woman, she was one of the first female scientists hired by Iowa State University. Spending time in the field with her on our annual Botany Club trips to the Utah desert, the usually formal Dr. T dressed down in blue jeans and a sweatshirt, loved to sing songs around a campfire, and didn't mind throwing her sleeping bag down on the big tarp with all the students so she, too, could sleep out under the stars. On the first trip to Natural Bridges National Monument in our yellow school bus, Dr. K and Dr. T guided twenty students to help catalogue plant and animal species before the park was officially opened to the public. We camped in desert hideaways, carting everything in and out past pinyon pine and juniper—the only trees around. There were no paved roads. As Dr. K drove the bus in and out of dry arroyos, he regaled us with wonderful stories. Dr. T organized all the meals, catalogued all the collections, and provided a good deal of the identifications. I remember once asking her to identify a beautiful yellow flower I had found in the bottom of one of the arroyos early one morning. She was delighted to let me know I'd discovered an evening primrose that would slowly close up its flower as the day brightened.

During my undergraduate years in the late 1960s and early 1970s, there was no discipline called plant energetics. There was plant physiology, plant anatomy, and plant ecology, but the term *plant energetics* was barely used. Now there is an entire field of plant energetics, with the currently definitive book on the subject titled *Plant Energetics*, written by a Ukrainian and a Russian. The book explains in detailed scientific language the quantum and thermodynamic processes that enable the conversion of solar energy by plants, including photosynthesis and transport of nutrients through the roots, trunk, and branches. The book also explains how understanding plant energetics is valuable to applied research in agriculture, horticulture, bioenergetics, biophysics, photobiology, and plant physiology.

Plant Energetics

My personal search to integrate my physical connection with trees to my intellectual comprehension of their botanical characteristics was timely. I had stumbled into my arboreal connection through experience and initiation at the same time that whole disciplines were being simultaneously developed in the scientific community.

About the time that my tiny kayak completed its circumnavigation of Lake Superior and touched the sand at its Park Point starting place back in Duluth in August 1992, Healing Touch was offering its third year of instruction, and studies of plant energetics were being introduced into scientific journals. Healing energy modalities, such as Reiki and Chinese acupuncture, well-known and accepted in Eastern Asia for centuries, were gaining increasing exposure in Europe and North America. But most Westerners, myself included, had been raised to understand things through the lens of science. So, having no intellectual explanation for why I kept reacting to trees as I did, I kept pursuing more information. And the more I learned, the more I understood the commonalities between trees and other life-forms—including myself.

In my experience of cramping and hypothermia that cold June day so many years ago, my desperate call for help was responded to by the only other living system in my awareness—aspen trees. At the time I was too depleted to question *how* these trees were communicating instruction to me—I just listened because my life depended on it. Now I understand that I can always connect to universal energy. And I understand that trees stand as near perfect models of how to do that.

When relaxing against a tree's trunk, I have the opportunity to align my human energetics with the tree's energetics. This alignment is not science, not intellect. It is experience. As I acknowledge all that is going on within the tree's trunk, I become more conscious of parallel processes going on in my own body that are also drawing energy from the earth and the sun. The tree's presence and my awareness of the tree's presence remind me how to be more fully alive. Communication through awareness is how living systems relate: the tree and the ground, the tree and the birds in its branches, the bugs and the bark, the mammals in the forest seeking shelter under the overturned roots, the woman leaning against the tree.

Meeting a Tree

In a recent teaching experience, my partner, Christina Baldwin, and I led a group of Louisiana Methodist clergy and laity to witness an old cedar tree in the Olympic National Park. Three vans pulled into the empty parking lot along U.S. Highway 101 near Kalaloch in far western Washington. Midlife men and women climbed out of the vans, finishing conversations, looking for cameras and backpacks, and searching for the big tree they had come to see. Christina rang a bell, and everyone gathered around us.

⌃ *The old cedar tree on the west coast of Washington state.*

We were standing in a muddy parking lot surrounded by immense western red cedar and hemlock trees. The dense forest canopy coupled with the low-lying fog created a feeling of dusk instead of midday. There was no sound from the nearby highway. There were a few high call notes from a small flock of golden crowned kinglets flitting through the thick underbrush of salal, huckleberry, and salmonberry. The group quieted and awaited instruction. "I want to share a few things about the old cedar tree that lives about a half block on the other side of that path," I said, gesturing toward the underbrush behind the group.

"This tree is a true elder. It is likely different from any tree you have ever seen—more immense and more complex. You will have to really study to distinguish its trunk from the trunks of those growing on it for support. Everything about it speaks of interconnectedness."

At this point I dropped my voice lower so people had to lean in and listen more carefully. "The tree began growing over one thousand years ago. Slowly, surely, it escaped the grazing of deer and elk, it managed to survive wind and storm, and eventually it found its place in the upper reaches of this canopy. In its life it has produced tons of oxygen, has sheltered uncountable birds and mammals, and has been worshipped by Native peoples. You would be hard-pressed to find another creature that has contributed so generously to its neighbors.

"I invite you to remain silent during your visit with the old cedar. Stand a distance away and really study this great tree. Be sure to spend time looking at it from all directions, and don't hesitate to approach and touch it. We'll ring the bell when it's time for everyone to regather at the vans."

Fifteen of us entered the cathedral of the forest. No one spoke. The natural silence around us was palpable. A few people entered the cavern of the hollowed out trunk in the center of the tree. Some leaned against the outer trunk and peered up through hanging moss and epiphytes (plants that grow on other plants) seeking an elusive top. All of us were hushed by the presence of something greater than ourselves.

After a while a couple joined us, then a family. Our silence invited them to join our reverie. One tourist even said aloud, "It's kind of like church here." Several people near her simply nodded. The imprint of respectful silence was stunningly inspirational. For thirty minutes we changed the history of human–tree interactions.

Inspiration was the gift we carried out of the woods that morning. We said, "Wow, that tree was so inspiring." The tree was not doing anything different because we were there: We were being different because we were there. This is our job—to notice, to appreciate nature, to become inspired, and then to act. The tree stands its ground and offers itself to our experience, and if we show up, we become keepers of the trees.

We're All in This Together

A recent storm in the Puget Sound region reminded me again of our deep and fragile connection with trees. For three days the media predicted a dangerous approaching storm. My VHF marine radio forecast heavy rains, flooding, and sustained winds of thirty-five to forty-five miles per hour with gusts as high as eighty. Cliff Mass, the enthusiastic, well-known professor of atmospheric sciences at the University of Washington and local weather guru, warned of conditions paralleling the perfect storm. He suggested that people with south-facing bedrooms bordered by trees sleep in other parts of the house on this mid-December night.

Rain began falling by mid-morning—not the usual steady, light Northwest winter drizzle, but pounding rains that filled up storm sewers and gutters, overloaded drainage systems, and sent rivers over their banks. The brooding, ominous sky darkened and lowered as the day progressed. By day's end there was standing water in all the low spots in our yard, and the predicted wind finale was beginning to rev up. Donning a rain jacket and rain pants, I headed out into the dusk to visit each of the tall trees bordering our little rambler. These Douglas firs shelter us from the frequent winds blowing off Puget Sound, and in big storms like this Puget Sound convergence, our safety and shelter depend on their strength. This storm would test them. I immediately headed to the old grandmother tree behind our home.

Laying my cheek against the furrowed bark, I felt its familiar strength. Despite the wild thrashing of branches overhead, I could feel no movement in the four-foot-diameter trunk of this hundred-year-old fir. (Perceived rocking or swaying at the base of a tree indicates its roots are not holding.) In a voice loud enough to hear myself above the terrifying roar of the wind, I said, "Stand firm, my friend. My prayer is for no damage to you or our property."

Dashing thirty feet to the next tree, I repeated my hug, test of the roots, and offered blessing. This was a ritual both of assessment and of assurance: an amalgam of the scientifically trained botanist and the longtime lover of trees. By the time I had finished the pilgrimage to each of our nine trees, it was no longer safe to be outside. A steady barrage of cones and small branches was bombarding me, and the loud snaps of large branches succumbing to the storm were more frequent overhead.

Most of that night the wind howled up the street, screaming through the tiny gaps in our slider windows, sometimes literally shaking the house. By 8:00 PM the electricity was out. We lit oil lamps and huddled around the wood stove in the living room on the north end of the house. We spent a restless night sleeping on the couch. And even I, who loves the wildness of storms, wished that the wind would stop.

By morning light the carnage was evident. Huge limbs littered the yard—some were twenty-feet long and three to four inches in diameter at the base. Miraculously all nine of our trees were still standing, and no damage had been done to either our home or our small backyard office. Many other people were not so fortunate. The mighty trees of the Northwest, standing in waterlogged soils after a month of record rainfall, fell through homes and garages, crushed cars and fences, and knocked out power for 1.5 million people. Utility crews called in from a four-state area worked around the clock clearing trees off power lines. We got our power back in three days, but thousands of people in the Seattle metro area did not have electricity again until Christmas Eve—nine days later.

The day after this huge winter storm in the Pacific Northwest, our little corgi dog, Glory, and I ventured out to sit beneath the largest of the Douglas firs in our yard and

Ann and her little dog the day after a huge Puget Sound storm. »

⌃ *Ann with her grandson, Jaden.*

survey the damage. I squatted and nestled my back into the wet, furrowed bark of our oldest tree. Glory sat between my legs and leaned her little brown and white furry body into mine. Together we observed the morning. The trees in and around our yard stood miraculously still after surviving the gale.

Glory's little pointed ears twitched and rotated, and soon I, too, heard the high-pitched sounds of tiny golden-crowned kinglets coming in to drink at the birdbath near us. With their gold and black striped crowns and pale brown and green bodies, they look like fairy royalty. My dog turned her gaze and remained perfectly still watching them dart from the branch next to us down to the bath and back to the branch. Life was resuming after the great storm. Rising to begin the long task of clearing the yard, I stood knee-deep in the debris of these strong sentinels, glad beyond gladness that we had all made it through the night.

Nearly fifteen years earlier, a small grove of aspens had told me to breathe. They could not jump across the interspecies barrier to help me in any way except presence. I believed them. I made it through. Now I had told a grove of tall Doug firs, "Hold on." I could not help them in anyway, except encouragement. Did my prayer help? I don't know. We made it through.

And that's the point: Trees and people need to make it through together.

> A percentage of the proceeds from this book will be distributed to various keepers who have so generously shared their stories—those whose work is ongoing and who need public support.
>
> Certainly you are invited to join me in watering their roots.
>
> Contact information is provided in the resources section.

Photo Courtesy: Ann Linnea

Photo Courtesy: Ann Linnea

Notes

CHAPTER 1: The Tree Man of Los Angeles

TreePeople's Web site, www.treepeople.org, contains a wealth of information and connections to many of the projects they are involved in.

Lipkis, Andy and Katie Lipkis, with TreePeople. *The Simple Act of Planting a Tree.* Los Angeles: Jeremy P. Tarcher, Inc., 1990.

CHAPTER 2: The Steward of Wildwood

Niosi, Goody. *Magnificently Unrepentant.* Surrey, BC: Heritage House, 2001.

Loomis, Ruth. *Wildwood: A Forest for the Future.* Gabriola, BC: Reflections, 1990.

Pp. 102–27, "Pacific Suite," Douglas Chadwick, *National Geographic*, February 2003.

CHAPTER 3: The Tree Pruner

International Society of Arboriculture, www.isa-arbor.com

Pruning facts, various agriculture extension brochures, and Web sites

CHAPTER 4: The Man Who Turns Wood

Much of the research about rain forests in this chapter was conducted on the Web because the issues are changing so fast. Useful sites include the following:

- Mongabay.com, http://news.mongabay.com, is a popular environmental science and conservation news site.
- Greenpeace, www.greenpeace.org, is a longtime environmental action group.
- Rain Forest Alliance, www.rainforest-alliance.org, is dedicated to the conservation of tropical forests for the benefit of the global community.
- *Science* magazine, www.sciencemag.org, specifically an article from *Science*, October 23, 2009, Vol. 326, no. 5952, pp. 516–17, "Green Energy: Another Biofuels Drawback; The Demand for Irrigation," by Robert F. Service. And from that same issue is the Policy Forum Climate Change multiauthored paper, "Fixing a Critical Climate Accounting Error."
- PeaceTrees Vietnam, www.peacetreesvietnam.org, is a bomb-removal organization operating in the heart of the former DMZ in Vietnam.

« *Fall ginko leaves*

CHAPTER 5: The Tree "Doctor"

Quote from p. 316, Margaret Lynn Brown. *The Wild East: A Biography of the Great Smoky Mountains.* University of Florida Press, 2000.

Numerous Web sites were helpful in supplementing Kris Johnson's considerable expertise:

www.na.fs.fed.us/fhp/hwa/, www.nps.gov/grsm/parkmgmt/index.htm (the management page of the Great Smoky Mountains National Park), and www.friendsofthesmokies.org/# (under project highlights and milestones)

CHAPTER 6: The Arborist and Big Tree Hunter

Eastern Native Tree Society Web site, www.nativetreesociety.org, is a comprehensive Web site with a search engine that connects to many, many research projects including the Tsuga Search Project.

CHAPTER 7: The Community Organizer

The Trees Forever Web site, www.treesforever.org/, contains extensive references on the numerous projects the group is engaged in.

The Alliance for Community Trees, http://actrees.org/site/index.php, was another important source for this chapter.

CHAPTER 8: The Weekend TreeKeeper

Within the Chicago Openlands Project, www.openlands.org/, various TreeKeeper projects are listed.

The Morton Arboretum, www.mortonarb.org

Chicago Park District, www.chicagoparkdistrict.com

Forest Preserve District of Cook County, www.fpdcc.com

Louv, Richard. *Last Child in the Woods: Saving Our Children from Nature-Deficit Disorder.* Chapel Hill, NC: Algonquin Books of Chapel Hill, 2005.

Carson, Rachel. *The Sense of Wonder.* New York: Harper & Row. 1956.

CHAPTER 9: The Businessman Who Loves Trees

Waverly Light and Power, http://wlp.waverlyia.com

American Public Power Association, www.appanet.org

Bill McKibben's international 350 project to stop global warming, www.350.org

The U.S. Department of Energy, www.energy.gov, provides much information on carbon sequestration, energy efficiency, and energy sources.

Plains Justice, http://plainsjustice.org/, is a Great Plains public interest law firm.

Energy Center of Wisconsin, http://www.ecw.org, for Web access to the *Energy Efficiency Guidebook for Public Power Communities*.

World Wildlife Fund PowerSwitch!, http://www.worldwildlife.org/climate/projects/ps.cfm

CHAPTER 10: The Wilderness Researcher

Public Law 88-577, 88th Congress, Section 2.a

Public Law 88-577, 88th Congress, Section 2.c

The WILD Foundation, www.wild.org/

Dawson, Chad P. and John C. Hendee. *Wilderness Management, 4th Edition: Stewardship and Protection of Resources and Values*. Golden, CO: Fulcrum Press, 2009.

Sharpe, Grand W., John C. Hendee, and Wenonah F. Sharpe. *Introduction to Forests and Renewable Resources: 7th Edition*. McGraw-Hill; reissued Long Grove, IL: Waveland Press, 2009.

U.S. Forest Service, www.fs.fed.us

Visit Treesearch, www.treesearch.fs.fed.us/pubs, for a list of all U.S. Forest Service research publications.

CHAPTER 11: The Man Called "Mr. Tree"

"Quantifying and Visualizing Canopy Structure in Tall Forests: Methods and a Case Study," by Robert Van Pelt, Stephen C. Sillett, and Nalini M. Nadkarni, Chapter 3, *Forest Canopies*, by Margaret Lowman and H. Bruce Rinker. Massachusetts: Elsevier Academic Press, 2004.

Van Pelt, Robert. *Champion Trees of Washington State*. University of Washington Press, 1996.

Van Pelt, Robert. *Forest Giants of the Pacific Coast*. Vancouver, San Francisco, Seattle: Global Forest Society and the University of Washington Press, 2001.

Save the Redwoods League, www.savetheredwoods.org

Pp. 28–63, "Redwoods the Super Trees," Joel K. Bourne, *National Geographic*, October 2009.

CHAPTER 12: The Activist Pruner of Emerald City

Turnbull, Cass. *Cass Turnbull's Guide to Pruning: What, When, Where and How to Prune for a More Beautiful Garden*. Seattle, WA: Sasquatch Books, 2004.

PlantAmnesty Web site, www.plantamnesty.org, contains information for obtaining the group's numerous helpful pamphlets.

Dr. Alex L. Shigo Web site, www.shigoandtrees.com

CHAPTER 13: The Native Carver

Murray, Liz and Colin Murray. *The Celtic Tree Oracle: A System of Divination*. New York: St. Martin's Press, 1988.

Back issues of the *Ashland Daily Tidings*

Agnes Baker Pilgrim Web site, www.agnesbakerpilgrim.org

Russell Beebe Web site, www.russell-beebe.com

CHAPTER 14: The Botanist Grandmother

Linnea, Ann. *Deep Water Passage: A Spiritual Journey at Midlife*. New York: Pocketbook, 1995.

Lachecki, Marina, Ann Linnea, Joseph Passineau, and Paul Treuer. *Teaching Kids to Love the Earth*. University of Minnesota Press, 1991.

Octavian S. Ksenzhek, Ukrainian State University of Chemical Technology at Dniepropetrovsk, and Alexander G. Volkov, University of California at Los Angeles, *Plant Energetics*, 1998.

Ann Linnea Web site, www.peerspirit.com

Photo Courtesy: Ann Linnea

Recommended Reading

*K*eepers of the Trees is a book based in story, science, and advocacy. It brings forward the total weave of people's lives as they became keepers of the trees. This suggested reading list is an eclectic reflection of the book's philosophy;

Baldwin, Christina. *Seven Whispers: Spiritual Practice for Times Like These.* Novato, California: New World Library, 2002.

Baldwin, Christina. *Storycatcher: Making Sense of Our Lives through the Power and Practice of Story.* Novato, California: New World Library, 2005.

Balong, James. *Trees: A New Version of the American Forest.* New York: Barnes & Noble, 2004.

Brown, Margaret Lynn. *The Wild East: A Biography of the Great Smoky Mountains.* University Press of Florida, 2000.

Carson, Rachel. *The Sense of Wonder.* New York: Harper & Row, 1956.

Dawson, Chad P. and John C. Hendee. *Wilderness Management, 4th Edition: Stewardship and Protection of Resources and Values.* Golden, Colorado: Fulcrum Press, 2009.

Freinkel, Susan. *American Chestnut: The Life, Death, and Rebirth of a Perfect Tree.* University of California Press, 2007.

Giono, Jean. *The Man Who Planted Trees.* Vermont: Chelsea Green Publishing, 2005.

Haupt, Lyanda Lynn. *Crow Planet: Essential Wisdom from the Urban Wilderness.* New York: Little, Brown and Company, 2009.

Hill, Julia Butterfly. *The Legacy of Luna: The Story of a Tree, a Woman and the Struggle to Save the Redwoods.* San Francisco: HarperCollins, 2000.

Johnson, Trebbe. *The World is a Waiting Lover: Desire and the Quest for the Beloved.* Novato, CA: New World Library, 2005.

Kaza, Stephanie. *The Attentive Heart: Conversations with Trees.* Boston: Shambhala, 1996.

« Winter alder silhouettes

Kernan, Sean. *Among Trees*. New York: Artisan, a division of Workman Publishing, Inc., 2003.

Kershner, Bruce and Robert T. Leverett. *The Sierra Club Guide to the Ancient Forests of the Northeast*. San Francisco: Sierra Club Books, 2004.

Kingsolver, Barbara. *Prodigal Summer: A Novel*. New York: HarperCollins, 2000.

Lachecki, Marina, Ann Linnea, Joseph Passineau, and Paul Treuer. *Teaching Kids to Love the Earth*. University of Minnesota Press, 1991.

Lanner, Ron. *The Bristlecone Book: A Natural History of the World's Oldest Trees*. Missoula, Montana: Mountain Press, 2007.

Linnea, Ann. *Deep Water Passage: A Spiritual Journey at Midlife*. New York: Pocketbook, 1995.

Lipkis, Andy and Katie Lipkis, with TreePeople. *The Simple Act of Planting a Tree*. Los Angeles: Jeremy P. Tarcher, Inc., 1990.

Little, Charles E. *The Dying of the Trees: The Pandemic in America's Forests*. New York: Penguin Books, 1995.

Louv, Richard. *Last Child in the Woods: Saving Our Children from Nature-Deficit Disorder*. Chapel Hill, NC: Algonquin Books of Chapel Hill, 2005.

Maathai, Wangari. *The Green Belt Movement: Sharing the Approach and the Experience*. New York: Lantern Books, 2003.

Marquart, Debra. *The Horizontal World*: Growing Up Wild in the Middle of Nowhere. New York: Counterpoint Press, 2006.

Morsbach, Hans. *Common Sense Forestry*. Vermont: Chelsea Green Publishing, 2000.

Nadkarni, Nalini. *Between Earth and Sky: Our Intimate Connections to Trees*. University of California Press, 2008.

Niosi, Goody. *Magnificently Unrepentant*. Surrey, BC: Heritage House, 2001.

Pakenham, Thomas. *Meetings with Remarkable Trees*. New York: Random House, 1998.

Peterson, Brenda. *I Want to Be Left Behind: Finding Rapture Here on Earth*. New York: De Capo Press, 2010.

Preston, Richard. *The Wild Trees: A Story of Passion and Daring*. New York: Random House, 2007.

Sauer, Leslie Jones and Andropogon Associates. *The Once and Future Forest*. Washington, D.C.: Island Press, 1998.

Scott, Susan. *Healing with Nature*. New York: Helios Press, 2003.

Sibley, David Allen. *The Sibley Guide to Trees*. New York: Random House, 2009.

Suzuki, David and Wayne Grady. *Tree: A Life Story.* Vancouver, BC: Greystone Books, 2004.

Turnbull, Cass. *Cass Turnbull's Guide to Pruning: What, When, Where and How to Prune for a More Beautiful Garden.* Seattle, Washington: Sasquatch Books, 2004.

Vaillant, John. *The Golden Spruce: A True Story of Myth, Madness, and Greed.* New York: W. W. Norton, 2005.

Van Pelt, Robert. *Champion Trees of Washington State.* University of Washington Press, 1996.

Van Pelt, Robert. *Forest Giants of the Pacific Coast.* Vancouver, San Francisco, Seattle: Global Forest Society and the University of Washington Press, 2001.

Wilson, Edward O. *Biophilia.* Cambridge, MA: Harvard University Press, 1984.

Zwinger, Susan. *Stalking the Ice Dragon: A Journey Through Alaska and British Columbia.* University of Arizona Press, 1991.

Zwinger, Susan. *The Last Wild Edge: One Woman's Journey from the Arctic Circle, Yukon to the Olympic Peninsula, Washington.* Boulder, CO: Johnson Books, 1999.

Photo Courtesy: Ann Linnea

Tree-Keeping Suggestions

WITH ACCESS TO THE INTERNET, readers can easily search to find local programs and resources for better understanding of, tending to, and advocating for the trees around them. The resources below are cited examples that are replicated in thousands of variations. Act locally and look for tree-based activities when you travel.

1. Plant a Tree

There are thousands of community organizations that sponsor tree planting. These organizations, four of which are featured in this book, provide tools, education, and funding to local citizens. The Alliance for Community Trees (ACT) Web site, http://actrees.org, lists organizations in forty states and Canada that provide tree-planting opportunities. A special note of support to Friends of Trees in Portland, Oregon, www.friendsoftrees.org, who reached out during the writing of this book.

- *International:* Tree Canada, www.treecanada.ca, see reference in #4.
- *National:* The Arbor Day Foundation, www.arborday.org, has nearly a million members who planted 8.7 million trees in 2009.
- *State:* The Wisconsin Woodland Owners Association (WWOA), www.wisconsinwoodlands.org, has many educational seminars on planting and caring for trees.
- *Local:* Spruce Up Austin, www.spruceupaustin.austincoc.com/, in Austin, Minnesota, is a classic example of a local organization spearheaded by the enthusiasm of one person, Mike Ruzek. In the group's twenty years of existence, it has planted over 2,400 trees, bringing together people from all walks of life to beautify the city.

2. Adopt a Tree and Care for It

Find a tree in a local park or other public place and adopt it. Visit it in all twelve months of the year. Bring water or mulch. Clean up around it. Spend time beneath it. Talk to others about caring for it—like it's not a place to chain a dog, carve your name, and so on.

« *Cedar of Lebanon in Germany*

3. Participate in a Recreational Tree Climb

For those who have the urge to get a bird's-eye view of the world, there are guides who furnish all safety gear and instruction. The *New York Times* published an article on January 7, 2005, on recreational tree climbing. One of the resource people listed was Genevieve Summers, www.dancingwithtrees.com, who is certified by Tree Climbers International and even created a Girl Scout badge about tree climbing.

4. Join an Organization That Supports Trees

This will give you an opportunity to make a meaningful donation to the cause of re-greening North America, and you will receive literature that will help you become more educated. The following list provides a tiny look at the wealth and breadth of tree-related organizations. Thanks to Cristy West, www.spiritoftrees.org, for her helpfulness in providing suggestions for this list.

- Tree Canada, www.treecanada.ca, is a nonprofit organization that provides education, technical assistance, resources, and financial support through working partnerships to encourage Canadians to plant and care for trees in rural and urban areas.
- American Forests, www.americanforests.org, is the nation's oldest nonprofit citizen conservation organization. The group maintains the *National Register of Big Trees*, plants millions of trees each year, and has a host of educational materials.
- Forest Ethics, www.forestethics.org, has worked with communities and industry to secure the protection of more than sixty-five million acres.
- Forest Stewardship Council, www.fscus.org, formed as a result of the 1992 Earth Summit in Rio, was created to change the dialogue about and the practice of sustainable forestry worldwide.
- Forest Protection Portal, www.forests.org, provides forest protection news, information retrieval tools, blogs, alerts, and other services toward the purpose of ending deforestation and the preservation of old-growth forests.
- Rainforest Alliance, www.rainforest-alliance.org, is dedicated to the conservation of tropical forests for the benefit of the global community.
- Greenpeace, www.greenpeace.org, is a well-known activist organization with a large focus on using its creative confrontation techniques to focus public attention on forest issues like tropical deforestation and destruction of old-growth forests.
- International Society of Arboriculture, www.isa-arbor.com, maintains a Web site containing information on the care and development of worldwide healthy trees. Members include urban planners, forestry professionals, com-

mercial horticulturists, and educators. The Web site contains many fact sheets on insect control and storm damage, journal abstracts, and links to land-grant universities and cooperative extension programs.

- Friends of the Smokies, www.friendsofthesmokies.org, is a fine example of a local organization that focuses on education, research, and activism for a specific place—the Great Smoky Mountains National Park.
- Save the Redwoods League, www.savetheredwoods.org, is another organization that has provided tremendous leadership through research, education, and activism to preserve the few remaining redwood stands.

5. Enroll in Community Classes

These classes are offered through local nurseries, arboretums, nature centers, county extension services, community colleges, local organizations, and by private individuals. Whether you are interested in learning more about the proper care and pruning of your own trees, learning more about what is happening with forest care and preservation, or learning to identify tree species, there is an abundance of often free or very low-cost classes to help you expand your knowledge base.

For an example of the kinds of thoughtful classes that are available from creative individuals, check out the Web site of Jean Linville, PhD, ecoartist/arts educator, www.treeofferings.com.

6. Explore the Inspiration That Comes from Trees

If you are a writer or an artist looking for a boost in creativity, take a "listening" walk in a local forest. If you are a person struggling with some life issue, don the appropriate outerwear and head into the forest in a spirit of looking and listening. If you are a parent or grandparent of young children with cabin fever, dress everyone up in appropriate clothing and head outdoors for an adventure in the trees.

7. Find a Mentor

If you are a wood worker, a carver, a pruner, a woodlot owner, or a budding naturalist, find people in your community who have been practicing your craft of interest. Ask them if they have classes or if you could assist in exchange for watching them at work. The age-old way of learning has been through apprenticeship. Many of the people in *Keepers of the Trees* sought out the tutelage of someone they admired.

8. Become a Mentor for a Young Person

Whether you are a parent, grandparent, uncle, aunt, neighbor, teacher, or community member, there is a young person whose life could be greatly enhanced by increased exposure to the natural world. This does not require a huge amount of skill—just interest and willingness. Invite a neighborhood child to help you prune or mulch in the garden. Occasionally share your morning dog walk with a child. Join the Big Brothers Big Sisters program and offer to spend your weekly time outdoors in nature. Create a family outing in a local park—bring a picnic, plan a treasure hunt, collect leaves, build debris huts, and make collages out of natural objects. Invite some neighborhood children over to help rake leaves with the promise of a good lunch. The opportunities and the need for this kind of interaction are boundless.

9. Become an Activist

The time is now to become an activist on behalf of trees. Any of the above steps constitutes becoming an activist. Additional actions include approaching your local utility company to see what it is doing about the proper pruning of trees and energy incentives for reducing carbon emissions. Focus on reducing your own carbon emissions by minimizing auto trips to the store or taking a bus. Find out what your own place of work is doing to recycle or minimize paper usage. Or take the time to record seasonal observations and appreciations of your own trees or those in a nearby park.

Being an activist is a different journey for each person. It is a journey that this precious planet is counting on us to take. Begin somewhere. Take the next step. The journey can open a whole new world of learning.

Acknowledgments

I BOW TO EACH OF THE thirteen keepers who shared their story. It has been an honor to get to know them. Their helpfulness in ferreting out details, photos, and memories has made this book a true cocreation.

The folks at Skyhorse Publishing are a creative force. From moving the publication date forward to get this out before Arbor Day to doubling the number of photographs, they have held intention for this book to find its way into the larger world as a true flagship to raise consciousness about trees. A special thanks to my editor, Jennifer McCartney, who has been a complete joy to work with.

Numerous people stepped forward to contribute photos. Thanks to each of the keepers and their organizations who provided photos: Sheila Boeckman at Waverly Light and Power; American Public Power Association; Laurie Kaufman and Melinda Kelley at TreePeople; Richard Gomez from the County of Los Angeles, Department of Public Works; Ashley Green at Trees Forever; Great Smoky Mountains National Park; Will Blozan; Dr. John Hendee and Marilyn Foster Hendee; Laura Robin; and Dr. Robert Van Pelt. Thanks to friends and family: Joe Villarreal, Marcia Wiley, Jeanne Petrick, Marie Mannatt, Stormy Apgar of www.ThunderCloudImages.com, and Jim Young, www.hatpeople.com. They donated time and energy to bring this book alive visually through their photographs.

To PeerSpirit seminar participants who held the story of this book supportively—a huge thank you. Your encouragement kept me going in more ways than you will know. To friends and family who kept listening to the unfolding stories of the keepers and believing in this book—wow, you're the best!

My dear sister, Margaret Brown, and my longtime friend, Janelle Brown, provided wonderful editorial support through many phases of this book. Debbie Dix, PeerSpirit office manager, held together business details so I could disappear for weeks at a time to write, provided helpful editorial and layout feedback, and cheerfully stayed the course of optimism.

There are not enough words for "gratitude" in the English language to describe my appreciation for the tremendous support that my partner, Christina Baldwin, provided through all phases of this book. She brought her own fine editing skills, her deep love of the natural world, and her steady sense of humor in at all the right times. Christina, for you, the biggest bow of all.

About the Author

Photo Credit: Harriet Peterson

Ann Linnea
writer, teacher, facilitator, wilderness guide

*T*HOUGH HER ANCESTORS DESCENDED from trees several million years ago, author Ann Linnea has never quite stayed on the ground. Raised in Austin, Minnesota, and educated at Iowa State University, she first bonded with the sturdy oaks and maples of that region. Now residing on Whidbey Island in Puget Sound, she is a fierce lover of the fir, cedar, and hemlock of the temperate coniferous rain forests. Story goes that when looking to buy her current home, the realtor drove her into the driveway, she spied the incredible Doug fir towering in the yard and said, "I'll take it," before ever going in the house.

Ann is a woman wild at heart in the best possible way: a mother and nurturer, her first book, coauthored with friends, tended to the need to instill in children an understanding of nature. Written during her own childrearing years, the result is the award winning classic, *Teaching Kids to Love the Earth,* published in 1991 by the University of Minnesota Press. The book won the Benjamin Franklin Award in Best Parenting/Child Care, the Midwest Independent Publishers Association Award for Best Environmental Book, and was translated into Portuguese and featured at the Earth Summit in Rio de Janeiro.

In 1992, the death of her best friend catapulted her into longing for a significant midlife rite of passage. She designed a summer adventure circumnavigating Lake Superior's 1800-mile shoreline by kayak. It turned out to be the coldest, stormiest summer on record, a trip that took her to the edge of death and beyond her limits of endurance. She lived to tell the tale and wrote the acclaimed memoir, *Deep Water Passage, A Spiritual Journey at Mid-life,* first published in hardcover by Little, Brown, and then in trade paper by Pocketbooks. The book thrives by word of mouth and in book clubs.

Also published in 2010 is her co-authored book with Christina Baldwin, *The Circle Way, A Leader in Every Chair,* by Berrett-Koehler. This legacy book represents the culmination of fifteen years of pioneering work in the field of bringing circle process into the mainstream of cultural dialogue. Ann has taught the use of circle from corporate boardrooms to wilderness questing.

She has essays in a number of anthologies, most notably in *Arctic Refuge, A Circle of Testimony* (Milkweed Editions, 2001) which was presented to every member of Congress; *The Soul of Creativity, Insights into the Creative Process* (New World Library, 1999), *Gifts of the Wild, A Woman's Book of Adventure* (Seal Press, 1998), and *Another Wilderness, New Outdoor Writing by Women* (Seal Press 1994). In 2002 she was commissioned by a grant from the Governor's Office of Washington State to write a local natural and social history, *Journey through the Maxwelton Watershed.*

She is co-principal of an educational company, PeerSpirit, Inc., Life and Leadership through Circle, Quest, and Story (www.peerspirit.com). She teaches Writing Nature's Wisdom seminars, leads wilderness quests, and offers an illustrated lecture based on *Keepers of the Trees.* (www.keepersofthetrees.com)

Ann Linnea
Box 550 • Langley, WA 98260
360-331-3580 • linnea@peerspirit.com

Index

Advance Praise for Keepers of the Trees

Among the hazards of becoming involved with environmental issues are periodic bouts of frustration, righteous anger, and even depression. At these moments the best antidotes are a hike in the woods of your choice or Ann Linnea's *Keepers of the Trees*. Restores your faith that some of our most precious forests are in loving hands.

—Dr. Margaret Lynn Brown, author of *The Wild East, a Biography of the Great Smoky Mountains* and professor at Brevard College

This book couldn't be more timely and relevant. At a time when our leaders are struggling to find a balanced, workable solution to the global climate change problem, this book illuminates a way to effectively sequester carbon dioxide while enhancing the natural beauty of our surroundings. Community-owned public power systems have promoted the beneficial planting of trees for many years through the Tree Power Program, recognizing that trees provide shade to cool and lower consumers' electric bills. *Keepers of the Trees* provides valuable insights into how all of us can contribute to the greening of our surroundings.

—Mark Crisson, President and CEO of the American Public Power Association

A significant contribution to the human-nature connection. Ann Linnea profiles inspiring people, themselves inspired by forests and trees. The passions of the profiled are directly transmitted to the reader. I know Robert Van Pelt and Will Blozan personally and can attest to their exemplary roles as bonafide keepers of the trees.

—Robert T. Leverett, executive director, Eastern Native Tree Society; co-author of *The Sierra Club Guide to the Ancient Forests of the Northeast*

These stories all share a touching humility and deep appreciation of the value of trees to humans and their innate and intrinsic value as a part of many terrestrial ecosystems and the health of the planet itself.

—Dr. Chad P. Dawson, SUNY College of Environment and Forestry, co-author of *Wilderness Management, 4th Edition: Stewardship and Protection of Resources and Values*

Ann Linnea shows us that ordinary people can make a difference. In these daunting environmental times, she has written a book that inspires hope.

—Morgan Simmons, environmental writer for the *Knoxville News Sentinel*.

Ann Linnea's love for the natural world and for the people who have committed their lives to nurturing, sustaining, and protecting it, shines through every chapter like sunlight through the forest canopy. She is a master at translating scientific detail and illuminating the life work of the tree keepers into clear, down-to-earth language. Not only are these stories as absorbing to read as a good novel, they inspire us to recognize how vital our connection is to nature and to each other.

—Dr. Susan S. Scott, author of *Healing With Nature*

Fourteen vivid stories of heroes, the practical sort we so desperately need, combine good storytelling with fascinating, usable science. Readers will find it applicable to their own backyards as well as the forests around the world. This is also a dangerous book— you will find yourself defending urban trees against bad pruning or removal, fighting to save forests against exotic pests, and becoming weekend warriors and tree keepers. But not to worry, you will also learn how and why leaning against your favorite tree brings peace and healing.

—Susan Zwinger, author of *Stalking the Ice Dragon* and *The Last Wild Edge*

Inspirational snapshots of diverse people with a common thread— life transforming experiences due to each one's involvement with trees. Greening the economy and sustainability are the watchwords of the day, and *Keepers of the Trees* is a new look at the original "green." If you thought you knew trees, think again and read this book for a fresh and uplifting perspective on the timeless beauty and enduring value of the trees we may all take for granted.

—Jan Schori, counsel to Downey Brand attorneys LLP and former general manager, Sacramento Municipal Utility District

Ann Linnea brings a fresh and welcome voice to nature writing—introducing us to the world of trees through the lives of those dedicated to be keepers of these Standing People. An inspiring and generous book.

—Brenda Peterson, author of *I Want to Be Left Behind: Finding Rapture Here on Earth*

In *Keepers of the Trees,* we meet a gaggle of individualists whose devotion to trees leads them to do strange and wonderful things that challenge our complacency. Read this at your own risk for you may not be able to resist joining them in advocacy of Earth's large, woody vegetables.

—Dr. Ronald M. Lanner, author of *The Bristlecone Book*

Shows that tree lovers come from a variety of backgrounds, and their passion takes as many forms as a great oak has roots. This book affirms our individual perceptions of trees and gives us even more reasons to stand in awe of them.

— Trebbe Johnson, author of *The World Is a Waiting Lover*

A sensitive and thoughtful vision of America's most beautiful and threatened forests, and of the extraordinary people who dedicate their lives to exploring, protecting, and "rewilding" these forests. Linnea offers us great portraits of people I know and deeply admire, including Kristine Johnson of the National Park Service, Will Blozan, who's a discoverer of the largest trees in Eastern North America, and Robert Van Pelt, a tree-climbing scientist and artist whose drawings of giant trees have changed forever how we see these organisms.

— Richard Preston, author of *The Wild Trees* and *The Hot Zone.*

An excellent book for all who love trees and nature. Catures the essence of visionary thinkers like Merve Wilkinson, and the dedication of others who have devoted their lives to being in harmony with trees. A good and inspiring read.

— Bill Turner, executive director, The Land Conservancy of British Columbia

Even if your sole experience of trees is from the sway of a backyard hammock strung between pin oaks and sugar maples, or from the blur of grey pine and red firs viewed through car windows on a mountain drive, you will enjoy Ann Linnea's *Keepers of the Trees*, which invites readers to "jump across the interspecies barrier" into the lives and imperiled habitats of trees. In fourteen narratives that gracefully blend scientific inquiry and human story, *Keepers of the Trees* chronicles the paths of unique individuals who have found tangible and pragmatic ways to improve the environment through stewardship of trees. Why is this important? Only "the secret to life," as Linnea's former professor was known to shout excitedly in class— chlorophyll, carbon dioxide, oxygen. "The earth needs her lungs," Linnea concludes. The environmental architect, William McDonough, has noted that nothing of human design demonstrates the ingenuity and resourcefulness of trees; they "make oxygen, sequester carbon, fix nitrogen, distill water, provide habitat for hundreds of species, accrue solar energy as fuel, build complex sugars and fuels, make soils, change colors with the seasons, create microclimates and self-replicate." If trees are heroes of a sustainable planet, then the individuals profiled in *Keepers of the Trees* are models for how we all, through practical action, can become guardians of earth's most heroic workers.

–Debra Marquart, author of *The Horizontal World: Growing Up Wild in the Middle of Nowhere*